BREWING IN LINCOLNSHIRE

ADAM CARTWRIGHT

AMBERLEY

First published 2025

Amberley Publishing
The Hill, Stroud
Gloucestershire, GL5 4EP

www.amberley-books.com

All images are from the author's collection, unless otherwise stated.

ISBN 978 1 3981 2513 1 (print)
ISBN 978 1 3981 2514 8 (ebook)

British Library Cataloguing in Publication Data.
A catalogue record for this book is available from the British Library.

Typesetting by Hurix Digital, India.
Printed in the UK.

Appointed GPSR EU Representative: Easy Access System Europe Oü, 16879218
Address: Mustamäe tee 50, 10621, Tallinn, Estonia
Contact Details: gpsr.requests@easproject.com, +358 40 500 3575

Contents

Acknowledgements and a Note on Sources

The frequency of brewery takeovers means that the historian is lucky if much documentation relating to past brewing firms has survived. The stories of Soulby's and Melbourn's had to be reconstructed from newspaper reports, street directories and the records of other firms, for instance, and although the corporate records of Hewitt's and possibly Sergeant's have survived in the hands of the National Brewery Heritage Trust, they were inaccessible at the time this book was compiled. North Lincolnshire Archives, however, have a smaller collection of Hewitt's material. Some of Mowbray's company records survive at Lincolnshire Archives, and Norfolk Records Office has a number of books and records relating to Soames. The website of the Brewery History Society is a very valuable resource too.

The chapters on 8 Sail and Ferry Ales were based on interviews with their respective owners, Tony Pygott and John Cussons, who kindly gave up their time to help – many thanks, gentlemen.

I should also like to thank the following for their help with this book, especially regarding photographs: the Brewery History Society, Grantham Civic Society, Grimsby History Club, Horncastle History & Heritage Society, Richard Croft, Ruth Crook, Michael Elsden, Mat Fascione, Nigel Fisher, Roy Hackford, Paul Holmes, Tony Laverick, Stewart Marsh, Brian Mossemenear, Allan Senior, Mark Shirley, Ken Smith, Simon Wilkinson and Kev Woodward.

Introduction

Lincolnshire is a huge county, the second biggest in England, with much of the land given over to arable farming. A key ingredient in beer is malted barley, a cereal that Lincolnshire has grown huge quantities of for centuries, especially on the Wolds and the heathland in the west and north. Former malthouses associated with processing barley can be found across the county, some on an enormous scale and all now derelict or in alternative uses. Local spring water was once used too, and in more recent times we've even been able to grow hops commercially here. All this shows that Lincolnshire beer can often be very local indeed.

This book looks at just a representative spread of breweries from across this vast county, although you won't find Lincolnshire's most famous one here. Batemans of Wainfleet celebrated their 150th anniversary in 2024 and they have a whole book to themselves, *Good Honest Tales*, which was published in 2024. But you will find the story of eight other Lincolnshire breweries in this book. Five of them sadly rolled out their last barrels over fifty years ago; one historic brewery is still in operation and the other two are both modern developments, less than fifteen years old.

Pubs and the breweries that serve them have been up against some very bleak economic conditions for decades. Since 2000, a quarter of all pubs in the UK have been lost, leaving us with around 45,000, according to the British Beer and Pub Association – one for every 1,500 of us, which doesn't sound much. Lincolnshire has been badly hit: most villages are now without a pub, and some of those that remain have slimmed down their opening hours to weekends and maybe a couple of evenings a week, with meals the focus more than beer. Right now, there's little to suggest things will get much better for pubs. So far at least, breweries aren't showing a similar decline. At the end of 2023, there were 1,815 in the UK, only very slightly down on the previous year. But there have been some significant losses in recent times, many of them caused by financial pressures, made worse in some cases by the pandemic lockdowns. Microbreweries are still opening up though, as we'll see in this book.

Deciding on which firms to include was difficult. I have hopefully achieved some form of geographical balance, but if your town or favourite brewery isn't mentioned,

I can only apologise and offer the excuse that I couldn't cover everywhere in this book. I should also say that good photographs of old breweries are very difficult to find, so you'll find a smattering of pubs in here too. How brewery firms developed and improved their pubs is a key part of their story.

The breweries that I've included are mostly the larger ones, ones that folk may remember in some cases, but a few more modest firms have crept in, as have those I felt were the most interesting. So, if you'd like to know why two aristocrats took over a creaky old Victorian brewery in the 1970s, or what it might have been like to take part in a brewery auction in 1905, or how you can turn an old cowshed into a brewhouse, then read on.

Brewing is a fascinating and highly skilled trade, one which demands hard work for sometimes only moderate returns, but the end result – the pint in your hand – is always worth it. I hope I've done the story of these iconic breweries justice.

1

Hewitt's of Grimsby

For many years Lincolnshire's largest brewery, the story of Hewitt's slow decline from successful family business to complete closure has a dark backdrop of corporate indifference, compounded by the eventual demolition of the brewery itself. Despite these misfortunes, Hewitt's are fondly recalled in Grimsby, and several of their former pubs still display distinctive reminders of the great days of brewing in the town.

There were brewers in Grimsby before Hewitt's. The earliest was probably John Garniss, an auctioneer who branched out into brewing around 1805, establishing his New Brewery 'in the front of the East Marsh, near the Hope and Anchor', a pub that still exists at the eastern end of Victoria Street. Garniss is usually credited with founding the brewery that became Hewitt's, and operated it for about ten years. The brewery then seems to have passed to John Hobson, who put it up for auction in April 1853, wishing to retire. At that point the business consisted of a brewery, a 30-quarter malthouse and kiln, situated on Pasture Street, and serving fifteen pubs, but it failed to find a buyer. Eight years passed before Hobson, by then aged sixty-nine, was finally able to retire, leasing the business to Edward Smith.

The 1860s were the decade in which Grimsby started to expand rapidly, the town's population more than doubling as its port quickly grew in importance. Also growing was the demand for beer to slake the thirst of fishermen and sailors, something which hadn't escaped the notice of an ambitious young man called William Taylor Hewitt. Born at Tuxford, north Nottinghamshire, in 1833, the son of a successful draper, he initially went into business with his brother, Thomas, as fellmongers (dealers in animal hides), but by the time he was thirty William had become a wine merchant at Bawtry, a prosperous market town on the old Great North Road just over the border into Yorkshire. He'd also bought a couple of pubs, the Black Swan in Kirton Lindsey, Lincolnshire, and the huge Butcher's Arms in Laneham, near the banks of the River Trent in western Nottinghamshire, and leased the Angel Inn at Bawtry. These three were the first of many hundreds of pubs operated by Hewitt Brothers.

Meanwhile, over at Grimsby, Edward Smith was keen to divest himself of the Pasture Street brewery. The news reached William Hewitt over in Bawtry, who brought his brother Thomas back into partnership. Between them, they made an offer, and a

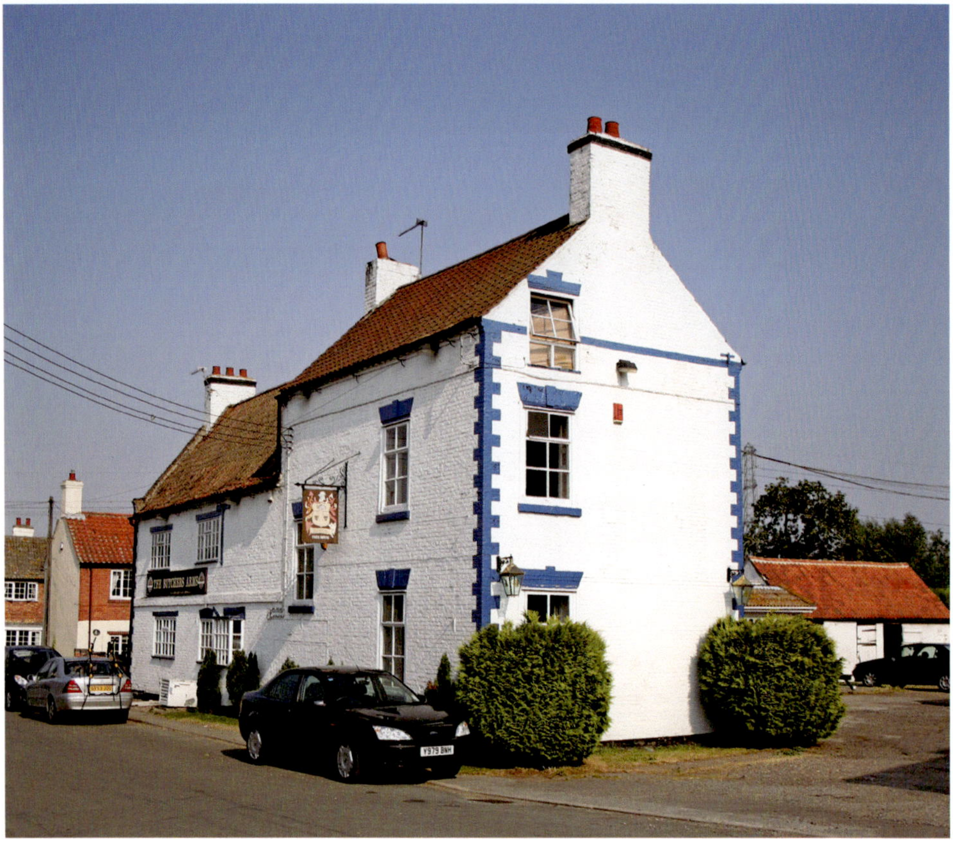

Just across the Nottinghamshire border was the Butchers Arms at Laneham, a very early Hewitt's pub. It closed shortly after this 2006 view and has since been demolished. (Richard Croft)

deal was agreed in 1869, making them the new owners of the brewery. It was a very bold step, partly financed by the sale of William Hewitt's entire stock of wines and spirits – just over 8,000 bottles at the Bawtry premises and imported cases and casks stored in a customs warehouse in London. William and his young family moved to No. 6 Pasture Street, Grimsby, a converted cottage next to the brewery, whilst Thomas remained at Bawtry to look after his many other investments; several advertisements of the time give the business address as 'Bawtry and Grimsby'. The Hewitt brothers also passed on their knowledge to their younger siblings; Edwin Hewitt was involved with the Exchequergate Brewery in Doncaster during the 1870s, moving on to a brewery at Yeadon, near Leeds. John Henry Hewitt was rather more successful, running the Ordsall Brewery at Retford from 1874 (where he became a long-serving councillor and mayor) and then taking on the Trent Brewery in Gainsborough, and more briefly the Brookhill Brewery in St Philips Road, Sheffield. Thomas Hewitt was a partner in both the Doncaster and Retford breweries.

William and Thomas Hewitt rapidly expanded the number of pubs served by their new concern, at first taking on some rather dubious establishments in Grimsby. It's

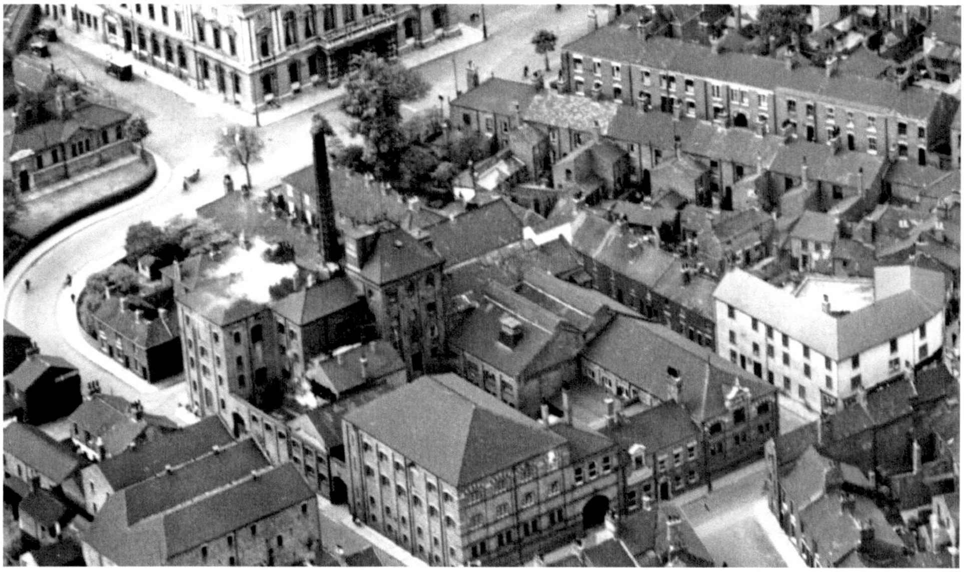

An aerial view of the entire brewery estate, with Pasture Street running left to right along the foot of the photograph. As a reference point, Grimsby's Town Hall can be seen at the top of the image and the surviving cooperage is to the left of the brewery buildings. (Brewery History Society)

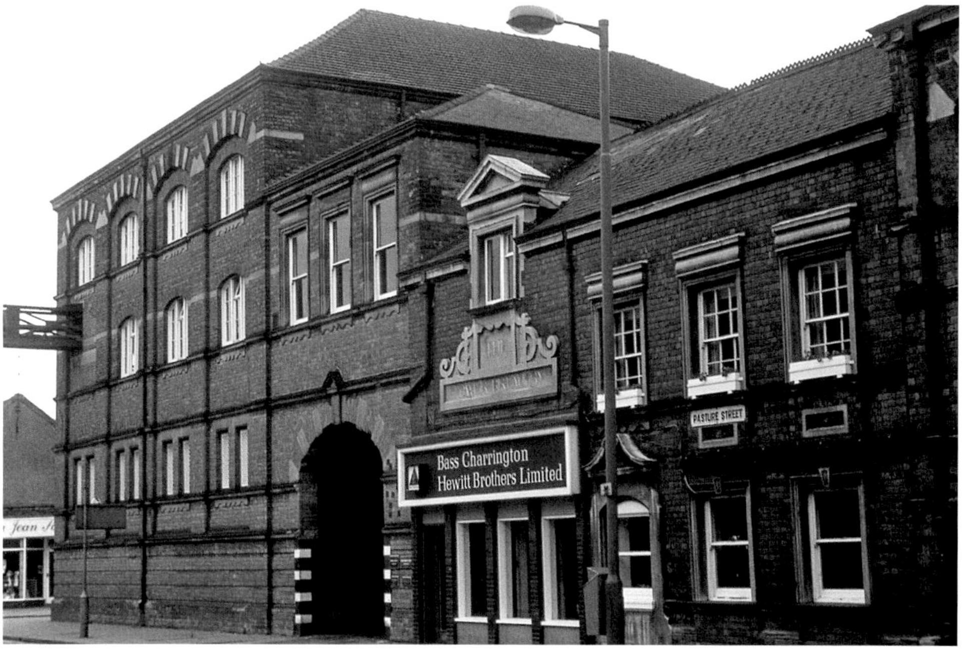

The impressive exterior of Hewitt's Tower Brewery in Pasture Street, Grimsby, during 1977. Nine years after brewing ceased, the premises had been demoted to offices and a distribution depot by Bass Charrington, although the signage carried the Hewitt name with equal prominence. (John Law)

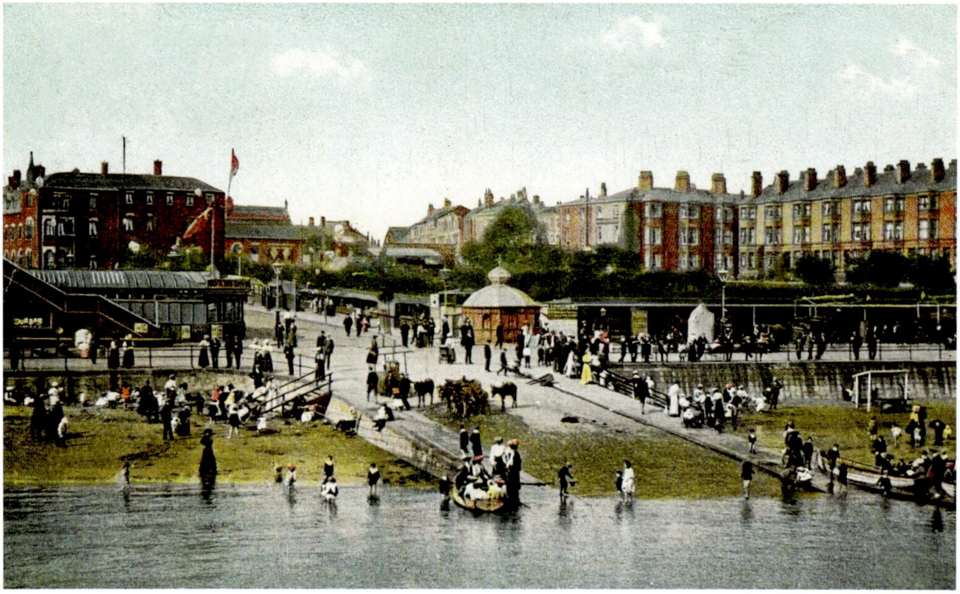

A postcard view of Cleethorpes promenade from around 1900. Hewitt's Dolphin Hotel was strategically placed by the seafront, on the far left of this bustling scene. (AC)

quite possible they were interested in making money as quickly as possible and, if so, it certainly worked. Legend has it that William visited as many local pubs as he could to persuade them to take Hewitt's beer. Many pubs still had what we'd now call microbreweries on the premises, probably in outhouses round the back. Maintaining quality would have been a struggle; William Hewitt would have told them he could solve at least that problem for them. Hewitt's had minimal serious competition in the area, enabling them to continue snapping up pubs with little opposition. An early example of their willingness to spend serious money was the old Dolphin Hotel on a commanding seafront position in Cleethorpes, which they expanded into a twenty-five-bed establishment in 1873. Apart from some jarring modern signage, the building remains largely unchanged externally. The brothers were also relaxed about distance, buying the Nag's Head and Prince of Wales beerhouses in Boston, almost 50 miles to the south, around the same time. Both were notorious old knocking shops which had probably been available on the cheap; the ease of supply by rail direct from Grimsby would also have made them attractive. Over time, Hewitt's started to weed out problem landlords. William Bickerton, the landlord of the Queen Inn, Barton-on-Humber, was allowing gambling and drunkenness on the premises. When he was charged by the police in May 1877, Hewitt's sacked him. On a lighter note, when Hewitt's bought the White Hart, an old coaching inn on Spilsby marketplace in 1886, they also found themselves lord of the manor, with the right to receive market fees and tolls – surely a first for a brewery. It was a pleasant change from unsuitable landlords and boozers of dubious reputations, especially as William Hewitt was climbing the ladder of Victorian respectability, becoming a Grimsby town councillor in 1877.

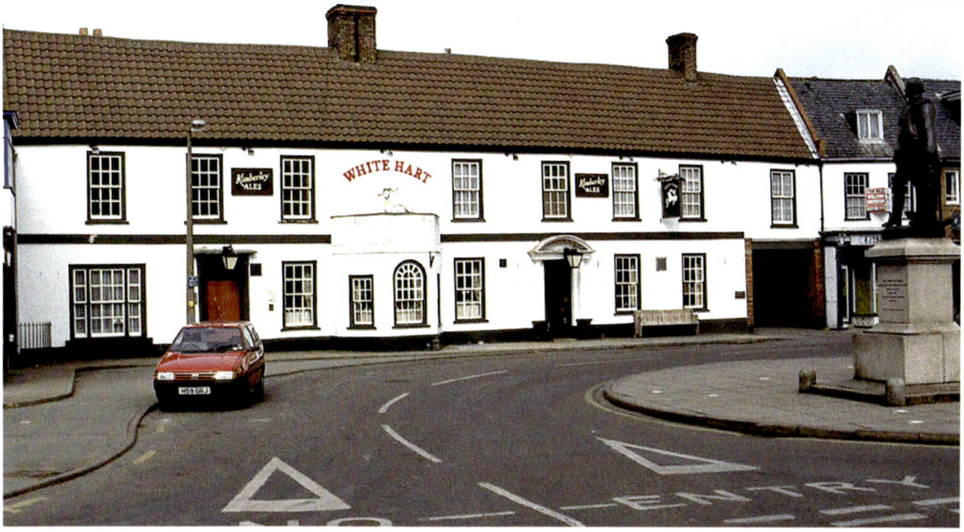

At the western end of Spilsby marketplace stands the White Hart, which Hewitt's acquired in 1886. By the time of this 1992 photograph it was owned by Hardy & Hansons of Kimberley, Nottinghamshire. Happily, you can still visit this attractive pub. (John Law)

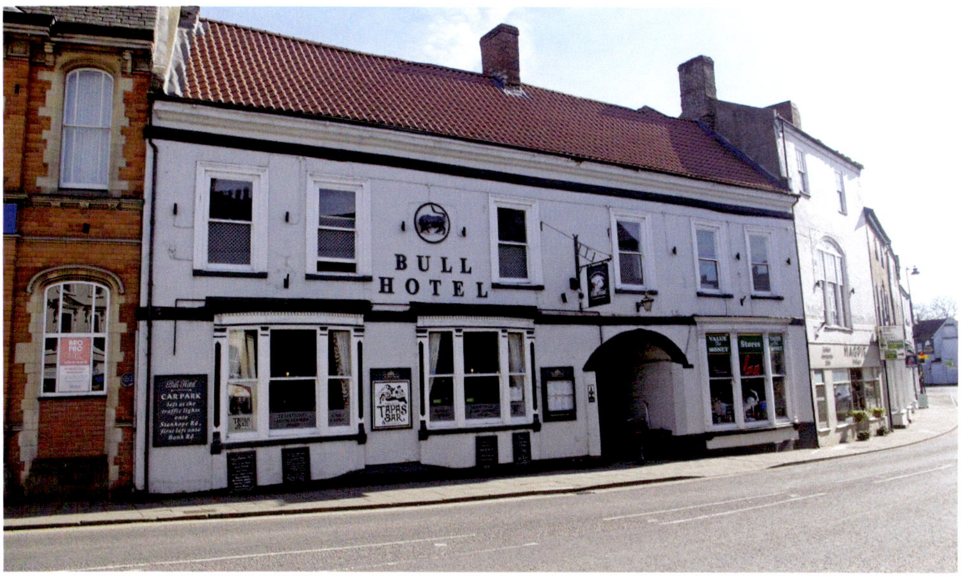

An old coaching inn, Horncastle's Bull Hotel has been a focus of town life for over 200 years. It's been owned by two of the breweries in this book. Mowbray's bought it in 1932 and in 1957 it was sold to Hewitt's in a package deal of pubs offloaded by Green's of Luton, who acquired Mowbray's a few years previously. The shop to the right of the carriage archway was once part of the pub but was de-licensed by Hewitt's a few months after their purchase. (AC)

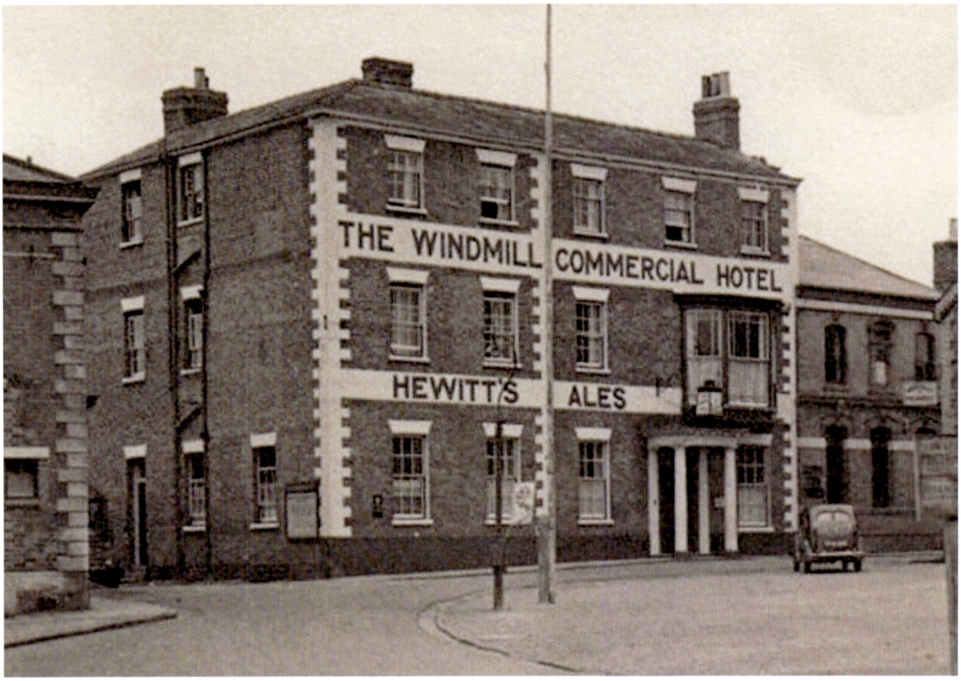

The Windmill Hotel in Alford remains a key feature of this small market town to this day. Bought by Hewitt's in 1888, there is little doubt as to ownership in this postcard view of the interwar years. Thomas Paine, author of *The Rights of Man* and often considered one of the founding fathers of the United States, was once a customs officer in Alford and had his office in the hotel. (AC collection)

The business now looked towards the next generation, with William's son, Thomas William Good Hewitt, joining the brewery in the early 1880s after studying at Bonn University. Like many aspiring young men from brewing families, he was probably sent to another firm to undertake a couple of years' training; that he patented two brewing inventions at the age of twenty-six strongly suggests that he understood the art of how to brew beer. Like his father, Thomas (usually known as Tom), was remembered as a hands-on businessman and employer, interested in all aspects of brewing, approachable and generous. He even insisted, it's said, on performing his own quality assurance, checking barley deliveries personally and tasting the brews.

In 1888, the business was incorporated as a private company, Hewitt Brothers Limited, with issued share capital of £200,000. All the preference shares were held by patriarch William Hewitt, with most of the ordinary shares divided between William and his son, Thomas. His other children, Titus and Mary, held a token forty shares each, suggesting they took no part in the business, but it's interesting to note that key employees were given a handful of shares too. Emmanual Carter, the head brewer, had five; the brewery manager, Herbert Pindar, had two, as did head clerk John Jowl; and maltster Thomas Johnson had one share. Not surprisingly, these men were long-serving employees. Noticeably, no other members of the Hewitt family were shareholders, but then at least one of them was extremely wealthy by this point: a few years earlier,

William Hewitt's brother, Thomas, had paid the enormous sum of £60,000 – about £8 million in today's prices – for the Finningley estate, over 2,000 acres of farmland just south of Bawtry.

The growing pub estate prompted enlargement of the brewery. By the late 1870s the Hewitts had moved out of the old house in Pasture Street to The Acres, a large new house on fashionable Bargate, on the new tramway route south out of town. It's likely that their old house was simply incorporated into the brewery complex, as suggested by surviving photographs. Other breweries were acquired too, including the Humber Brewery in Grimsby's Bull Ring (1892), the Gainsborough brewery of John Henry Hewitt (1893) and the Ordsall brewery in Retford (1897). But with such a cramped location, hemmed in by town centre streets, the only way Hewitt's could expand was away from the brewery site. They had already constructed a cooperage on East Street, opposite the brewery, for the construction, repair and storage of beer barrels (it's now the only surviving building from the old brewery), and adapted some adjacent houses as a barrel store. In 1900, they built a large new maltings a couple of hundred yards to the north on Victoria Street, designed by the Nottingham architect Richard Hardy, a brewery specialist. Though dwarfed by the massive maltings complex of Soames and Co. a little to the north (unrelated to the Spalding firm also covered in this book), Hewitt's were now able to malt their own barley – a key brewing ingredient – rather than buying in the finished product.

However, disaster would overtake the new maltings. One August night in 1901, a passing policeman noticed that an adjacent sawmill had caught fire, and within minutes it had spread to Hewitt's premises, which had been constructed as a traditional floor

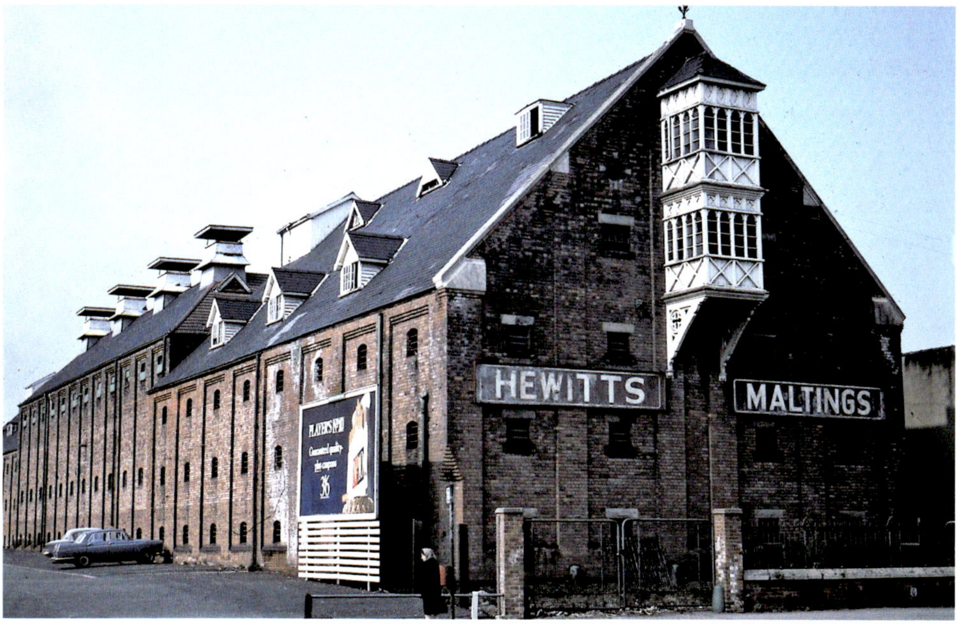

The impressive bulk of Hewitt's Maltings just off Grimsby's Victoria Street, seen in 1970, three years after closure. (Roy Denison via Brewery History Society)

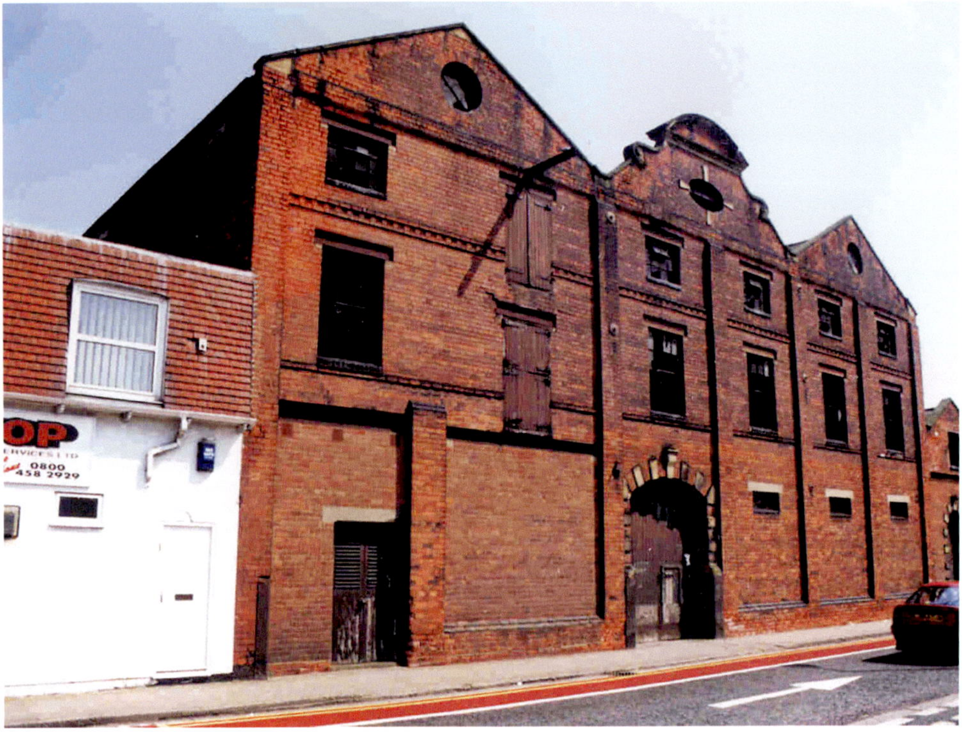

The cooperage in Town Hall Street, seen prior to renovation work which has seen it sensitively converted into luxury apartments. (Brewery History Society)

maltings with vast open floors on several levels. Such buildings were therefore highly vulnerable to fire, and so it proved. The malthouse was soon 'one mass of flame, which shot up to a great height' according to a local newspaper. Within an hour the roof of the malt kiln fell in, but the firemen gradually took control despite the salt water from the riverhead encrusting their hoses and valves, and by 4 a.m. the danger was over. Most of the maltings had been gutted, but the eastern end, with its fine timber balconies, was saved from serious damage. The major rebuilding required would have caused the business serious disruption.

A further blow came in April the following year with the death of the founder, William Hewitt, aged seventy-one, at Weelsby Old Hall, a vast old house on the southern outskirts of Grimsby that he'd bought some years previously. In his will he left the astonishing sum of £467,357, which roughly equates to about £64 million now, and clearly showed how successful the brewery had become. His eldest son, Tom Hewitt, became chairman and managing director. He inherited a company in good health, whose profits were often invested in the brewery and its tied estate, and buying more pubs as the company spread out into south and east Yorkshire. Some of Hewitt's most distinctive pubs date from this period, often featuring brightly glazed tiling to ground-floor exteriors, and always with the brewery's name prominently displayed; the legend 'Hewitt's Grimsby Ales' on many pubs proved more durable than the brewery itself, and can still be seen on a handful of surviving houses. The Plough, near

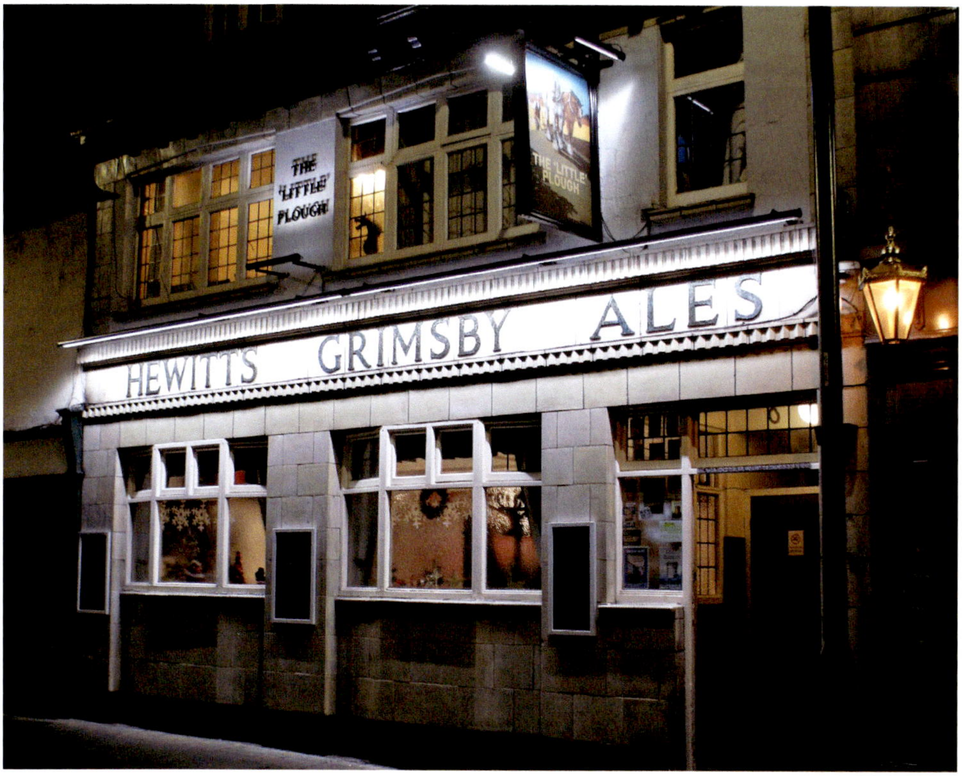

The Plough in West Laith Gate, Doncaster, is another 1934 rebuild, with even larger Hewitt's signage. Fortunately still open, it's usually referred to as 'the Little Plough'. (John Law)

Doncaster railway station, which retains a titled frontage incorporating the Hewitt's name, is a good example, albeit dating from the 1930s.

Like all brewers, Hewitt's were very seriously affected by the First World War. Beer production was heavily restricted by the government, compounded by most of the brewery's workforce being engaged on military service. The story of the Grimsby Chums is well known – a 'pals battalion' of volunteers, mainly from the town, of whom 810 did not return. Less celebrated, perhaps, is the town's Volunteers Training Corps, raised, financed and organised by Hewitt's long-serving general manager Harry Crosby, which provided reservists, who protected military installation at home. Tragically, Crosby shot himself in a Grimsby public lavatory in 1922, after thirty years working for Hewitt's; he was suffering business worries, exacerbated by wartime stresses.

Another figure from this time who gained some notoriety was Otto Overbeck, Hewitt's in-house chemist, a complex character and a lifelong bachelor. As the son of a prominent religious activist and linguist, he was home educated as part of an unconventional childhood. Equipped with a degree in chemistry, he moved into the brewing industry. As well as a brilliant polymath, he was an assured self-publicist, somewhat prone to exaggeration and hyperbole, meaning that accounts of the many interviews he gave to journalists need to be treated with some care. It's clear, though, that after spells at breweries in Wiltshire and Manchester, he arrived at Hewitt's in 1891 to take up a post as 'scientific director'. Like his contemporary Tom Hewitt, he'd studied at Bonn University and it's possible they had met there. 'Scientific director' sounds an impressive title, although it may have been Overbeckian elaboration: he was probably an elevated form of head brewer, responsible for beer quality and development. To be fair, some larger brewers had taken to employing chemists by this time; some even had laboratories, as they strove to achieve greater consistency in the final product; yeast in particular was vulnerable to infection. Overbeck's appointment certainly showed Hewitt's had a progressive and scientific approach to brewing.

Overbeck loved to experiment and show off his expertise. He devised a yeast-derived meat substitute called Carnose, no doubt a by-product of brewing; it sounded revolting and unsurprisingly failed to catch on, nor did a later attempt to feed it to cattle. But a later invention had more promise: non-alcoholic beer. The concept of low or no alcohol beer was not new; in fact, 'small beer', a low-strength beer, had been around for years. As an alternative to frequently polluted water supplies, the brewing process tended to kill off diseases. Overbeck's process was different, though: he removed the alcohol after the beer was brewed. His creation was supported by Hewitt's. Like all brewers, they were concerned about the influence of the temperance movement, particularly when their supporters popped up at licensing meetings pressing the magistrates not to renew licences. A willingness to produce de-alcoholised beer might tip the balance back towards the brewer. Hewitt's therefore agreed that part of their Ordsall brewery in Cobwell Road, Retford, could be used by Overbeck as a de-alcoholised beer factory; probably there was some surplus capacity there, but it's also likely they wanted Overbeck's venture kept well away from Grimsby.

The Overbeck promotion machine went into overdrive as he convinced several journalists that he'd been offered £50,000 for the rights to his process. He also delighted several hardened hacks with his life story ('The Wizard of the Brewery' – *Sheffield Independent*). However, he reckoned without J. R. Freeman of HM Customs & Excise's Newark office, who wrote to Overbeck in May 1914 telling him that under no circumstances could he operate a factory to produce de-alcoholised beer commercially, nor could he carry out any further 'experiments'. No doubt the Excise authorities viewed Overbeck's invention with horror: if his product became popular, huge amounts of excise duty, charged according to the percentage of alcohol in drinks, would be lost. Overbeck was furious and turned immediately to his pals in the press. Several newspapers printed his angry exchange of letters with Customs' head office in London, accompanied by a diatribe from Overbeck talking of 'a flagrant abuse of power ... absolutely illegal ... the sooner this is raised in Parliament the better'. He was eventually allowed to start production a year later, but by then it seems that both Overbeck and Hewitt's had lost interest in the project; the Ordsall brewery was quietly sold in 1935. He was, of course, decades ahead of his time; not until the hard-hitting road safety campaigns of the 1990s would drink-driving become largely unacceptable to many areas of society, combined with improvements in low-alcohol brewing technology.

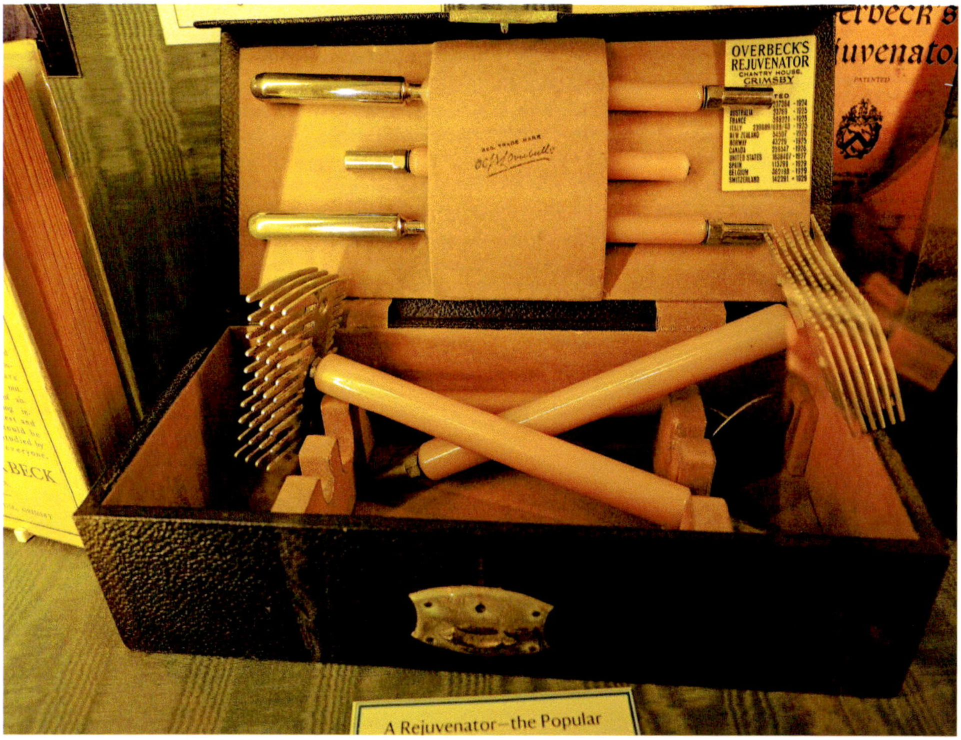

An example of Otto Overbeck's Rejuvenator seen here in its carry case. Rejuvenation was achieved by switching the machine on and applying the metal paddles to the owner's head. (Simon Wilkinson)

These setbacks couldn't keep Overbeck out of the picture, however, and in 1925 he hit the headlines again with his Rejuvenator, a fearsome-looking piece of kit consisting of two electrically charged paddles which resembled a pair of mallets. This new venture was born out of Overbeck's conviction that the human body was a form of electrical machine, whose various ailments could be treated by electrotherapy administered by the Rejuvenator. Overbeck, who was in his sixties and had left Hewitt's by this time, had lost none of his flair for publicity. His invention had transformed him from a dying man into 'a miracle of untiring energy, enough to tire out a whole party of sturdy youths', he told the press – a strange choice of words. But potential downfall was not far away; in 1928 he was charged with sexual offences involving teenage boys. Overbeck could afford to engage Norman Birkett KC to defend him, a criminal barrister at the height of his powers, who secured Overbeck's acquittal. Perhaps tellingly, he soon retired to Salcombe in South Devon, where he bought a large house and tropical garden paid for from sales of the Rejuvenator. He died in 1937, donating his home to the National Trust, who still maintain 'Overbeck's Garden' for visitors.

The death of the able Harry Crosby in 1922 had created a serious leadership vacuum at Hewitt's. By the 1920s it was clear that none of the third-generation Hewitts were interested in running a brewery. Tom Hewitt was ailing, unmarried and childless, and his nephews and nieces had more exotic interests, especially Titus Hewitt's son, Vivian,

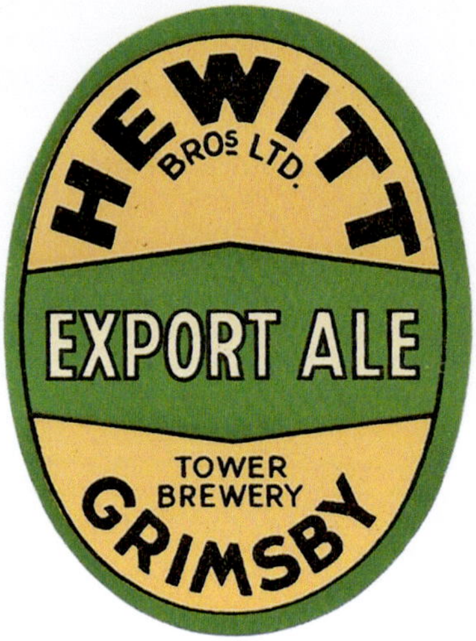

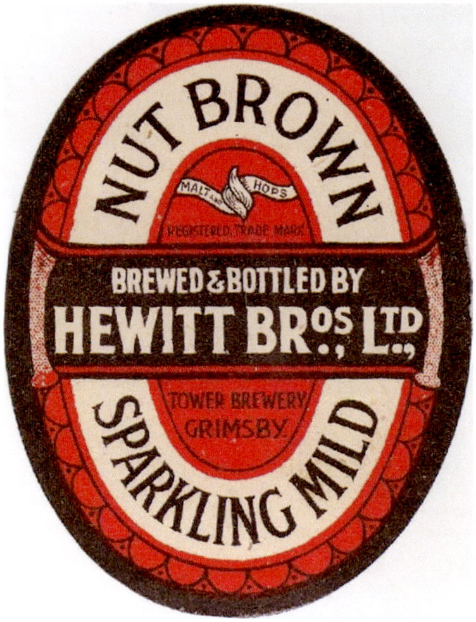

Above left: Hewitt's Export Ale label. (AC)

Above right: Hewitt's Nut Brown Sparkling Mild label. (AC)

an early aviator who later became a conservationist, albeit quite an eccentric one, in North Wales. Nonetheless, when Tom Hewitt died in 1930 at the age of seventy-three, Vivian Hewitt found that he'd inherited a large, highly profitable regional brewery and nearly 300 pubs. The business was being run in trust by the company's accountant Arnold Fowler, who Tom Hewitt had installed as managing director. In the eight years up to 1933, the company's net profits had exceeded £100,000 in all but one year. The trustees therefore decided to sell the company, with Vivian Hewitt's approval, but somehow news leaked out to the City and speculators got to work, scrabbling around trying to form syndicates to buy the highly attractive company. Confidential sale documents and accounting records were stolen from an intermediary by the simple process of getting him extremely drunk. It was a dark tale of fraudsters, undischarged bankrupts, cigar merchants, clandestine runners and schemes, most of which Hewitt's knew nothing about – although their representatives did meet with at least one of the rival groups.

Fortunately, these shady goings-on came to nothing. Hewitts' share offer was made public in January 1934, including the conversion of the business into a public limited company, with share capital of £500,000 and debenture loan stock of £700,000. Not surprisingly, the share issue was significantly over-subscribed. Sir James Calder, a highly respected City figure and a director of Wilson's brewery in Manchester, was installed as chairman, and Percy Chaplin, a wine merchant, joined Arnold Fowler as a director. The company was no longer in private hands and had to answer to external shareholders, including – ominously – a number of larger breweries, but at least it was able to resume its investment programme, installing new fermentation vessels at the brewery in 1934, and opening a new bottling store in 1937. Their new purchasing power also allowed them to snap up a couple of wine merchants in Doncaster and York, and five pubs and hotels in York, the furthest north they had ventured. The investment continued as new pubs replaced old. Examples were the Horse and Groom, the Hexthorpe House and the Stag in Doncaster, featuring much white tiling to the external walls, of 1933, 1934 and 1935 respectively and now all closed, and the Market Hotel in Yarborough Road, Grimsby, in 1937, a large but rather utilitarian two-storey corner property, with the same white tiling again and an 'HB' logo on the corner gable end that's still visible, despite the pub's conversion into a Tesco Express. It was followed by a new Humber Hotel on Grimsby's Humber Street, actually in the front garden of its predecessor (June 1938), again featuring much white tiling; unfortunately, it too is now in alternative use.

Slowly, Hewitts were becoming a large regional brewer, less rooted and less connected to the town of Grimsby. An indication of the future was the appointment in 1939 of Donald Burt as managing director, replacing Arnold Fisher. Burt came from Bass, the huge, internationally famous Burton-on-Trent brewers, where his father was a director. Hewitt's knew him well. Nine years earlier, when they were carrying out exploratory talks with bigger brewers about a sale, it was Burt who did the talking on

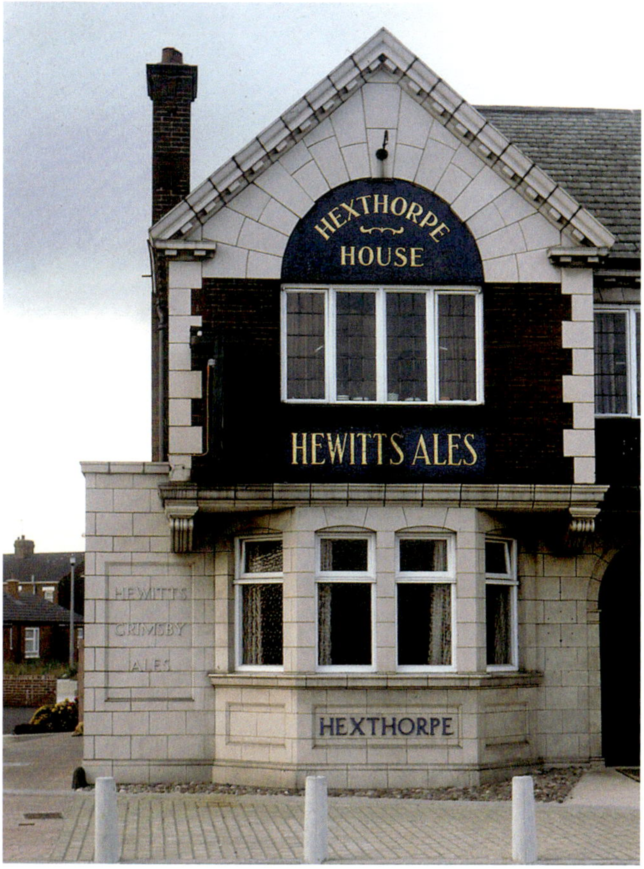

From Hewitt's 'white tiling' period: the Hexthorpe House, Urban Road, Doncaster, complete with a distinctive Hewitt's Ales sign which corporate ownership left unaltered. Now sadly demolished following closure in 2007, the pub was rebuilt by Hewitt's in 1934. (John Law)

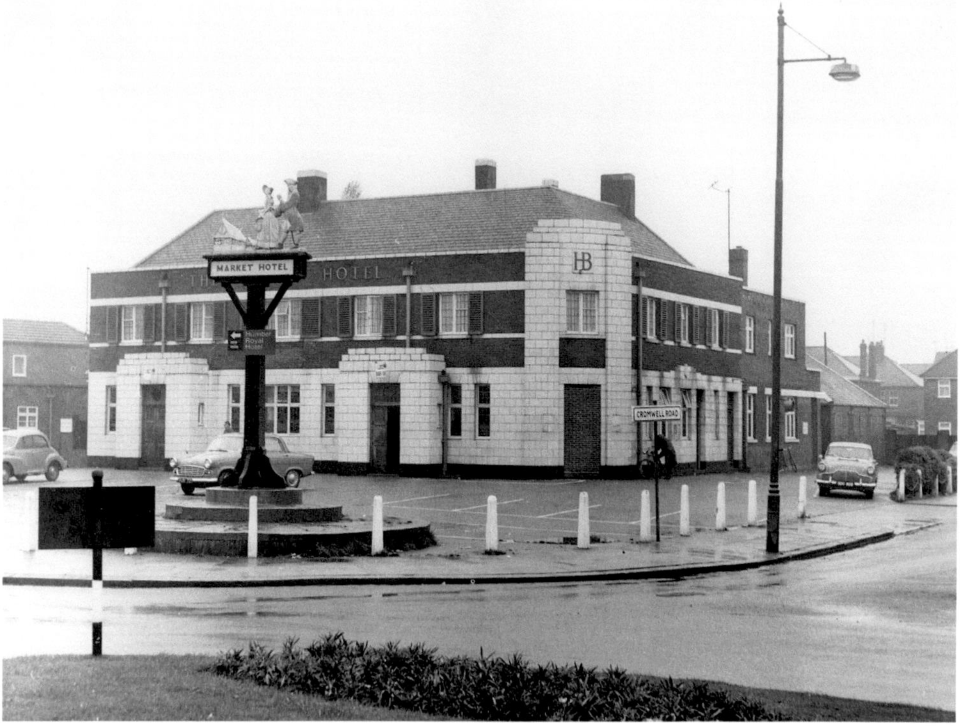

Another product of Hewitt's white tile period was the Market Hotel of 1936, a large building at the junction of Cromwell Road and Yarborough Road in Grimsby, seen here in the 1960s. The substantial wooden pub sign is impressive. Although the pub is now a Tesco Express, the 'HB' logo is still in place. (Grimsby History Club)

behalf of Bass. The firm were building up a shareholding in Hewitt's, as were other firms such as Whitbread, and no doubt Burt arrived at their insistence. For the time being, however, Hewitt's remained independent.

The Second World War again saw Grimsby under aerial attack, including the destruction in 1943 of Weelsby Old Hall, the former home of William and Tom Hewitt, as well as the Shades Inn, a tiny Hewitt's pub on Church Street, and an off-licence in Bath Street. Nearly 300 people died in bombing raids over Grimsby and Cleethorpes alone. Hewitt's themselves also lost the Peacock in Market Street, Gainsborough, bombed in 1942.

After the war, profits remained buoyant and the rebuilding programme resumed, this time mainly featuring completely new and much larger houses on largely post-war housing estates. The New Empire on Runswick Road, Grimsby, of 1956, was built jointly between Hewitt's and Bass, once again a clear indication of who called the shots; Bass ales also started appearing in some Hewitt houses, and in return Hewitt's bottled some Bass beers. By the end of the 1950s there were several more new pubs for Grimsby, as well as one each in Scunthorpe, Grimoldby, Swallow, Scartho and Gainsborough, and further new pubs for Hewitt's subsidiary Sergeant's of Brigg. A deal in 1956 with Flower's of Luton brought twenty-eight mostly town centre pubs across east Lincolnshire into the Hewitt fold. Those pubs were unwanted remnants of Soulby's of Alford, closed by Flowers (in their earlier J. W. Green identity) in 1952, and pushed Hewitt's pub estate past 300.

At the other end of the scale, it was closing time for older and smaller pubs, town centre and countryside alike, as their size or location, or both, made their returns unattractive. The Sloop at Tetney, on the Lincolnshire coast south of Grimsby, a haunt of anglers for some 200 years, closed one bleak February day in 1958 after years of declining sales. Like many rural pubs, it looked very much like a private house, with just a sign reading 'Sloop Inn Hewitt's Ales' to suggest otherwise. And pubs closed in Grimsby not just because they were small; the Chapmans Hotel on a street corner in the town's Central Market was in the way of road enlargement, so it closed in 1956. A particularly sad loss was the beautifully tiled frontage of the Lincoln Arms on Freeman Street, demolished in 1967.

The direction of travel towards Bass and Whitbread seemed to have been arrested in 1959 when a merchant bank called Hart, Son & Co. took control of Hewitt's. They had slowly been buying shares under the noses of the big brewers, driven by the determined leadership of Michael Richards and his investment vehicle Wood Hall Trust. Sir James Calder, the eighty-nine-year-old chairman, was replaced by Richards in May. Not quite a year later, it was announced in April 1960 that Hewitt's would 'merge' with a Hull brewery, Moors and Robson's, whose 140 pubs were nearly all clustered around eastern Yorkshire. One that wasn't was the Yarborough Hotel in Grimsby, a long-term rival of Hewitt's Royal Hotel; the deal would bring both under common ownership. In reality, it was a takeover by Hewitt's with the full approval of both boards, and Moors' Crown Brewery was immediately closed. In the turmoil that circled the British brewing industry during the 1950s, most friendly mergers took place in the hope of creating

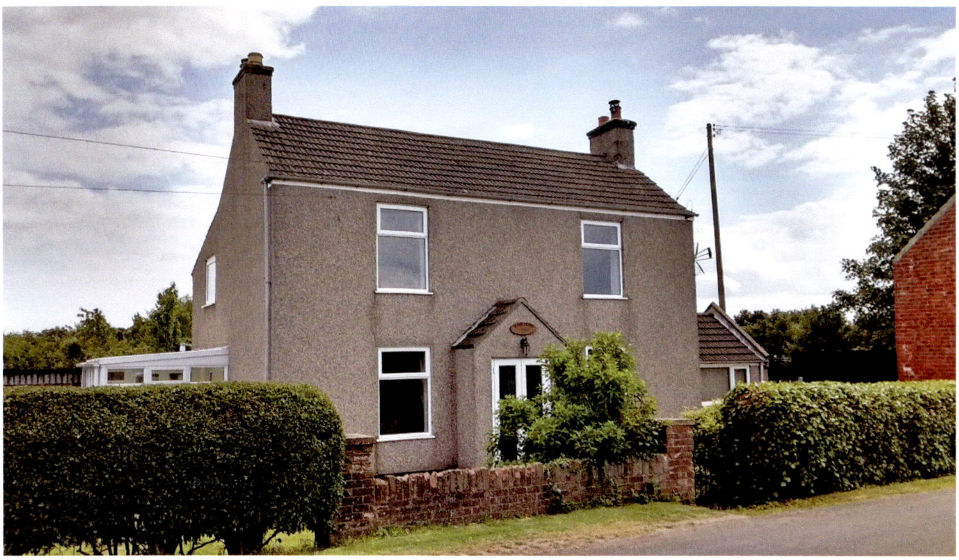

The former Sloop Inn at Tetney Lock, pictured in 2019. Despite the somewhat remote location, another pub, the Crown and Anchor, remains open about 100 yards away. (Stewart Marsh)

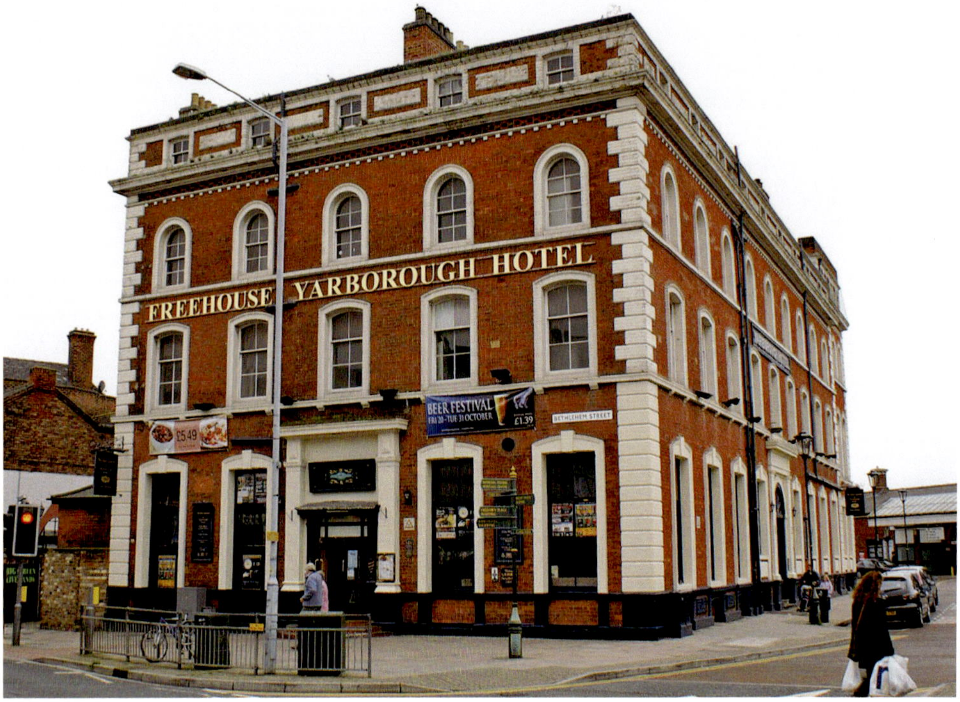

The Yarborough Hotel, a Grimsby landmark. Built for the Great Central Railway, Hewitt's acquired it when they merged with Moors' and Robson's of Hull. It's now a Wetherspoons hotel. (John Law)

a group that was unattractively large to corporate raiders, but such tactics usually didn't work out, if indeed that had ever been the intention in this case. The combined group supplied 480 pubs across eastern and northern England, but it was a short-lived triumph as large national brewing groups continued to form.

One such was United Breweries, formed in 1960 by a merger of breweries in Bradford, Sheffield and Edinburgh, but actually controlled by Canadian Breweries, founded by the industrialist E. P. Taylor. United made a £7.5 million bid for Hewitt's in December 1961 which was swiftly accepted, giving Wood Hall Trust an impressive profit of £1.8 million for its 39 per cent stake. With Hewitt's now controlled by United Breweries, and ultimately its Canadian parent, Donald Burt's departure was not long in coming; he retired in April 1962. A further merger with the London-based brewer Charrington created Charrington United Breweries later in 1962, a group with breweries spread across much of Great Britain and within which Hewitt's was now a relatively small component.

During the 1960s, the pub investment programme continued, with further new houses in Grimsby and others in Waltham, Cherry Willingham near Lincoln, Gainsborough, Immingham and Barton on Humber. Unfortunately changes in population and people's expectation from a pub have often proved terminal for large, open-plan establishments, many of which have not survived into the twenty-first century. The Mariners Rest in Grimsby's Albion Street, a marvel of electric pre-measured beer dispensing when it opened in 1963, became derelict for years until it was destroyed by arson in August 2024, although others of the period have survived in other commercial uses.

The Victoria Street maltings closed in 1967; floor malting in large Victorian buildings was increasingly being replaced by mechanical methods of processing barley. But the Pasture Street brewery had managed to stagger on during the 1960s, although increasingly its future looked uncertain as the large brewers pressed on with concentrating beer production into new modern brewing plants. Its fate was sealed – along with much of Hewitt's identity – when Charrington United Breweries merged with Bass in 1967 to create Bass Charrington, Britain's biggest brewer, serving almost 10,000 pubs. The hectic and bewildering tide of corporate mergers had finally brought Hewitt's into the Bass empire nearly forty years after the earlier discussions, but this time by association rather than intention. Rumours started to swirl that Grimsby's last remaining brewery would close, which were confirmed in a closure statement which cited cost pressures and the age of the brewery equipment. The last brew took place on 19 April 1968, with thirty staff losing their brewing jobs, although some were redeployed as the Pasture Street premises were retained as a distribution depot. Production was moved to Bass's brewery at Tadcaster in Yorkshire, from where beer was hauled over to Grimsby 'in special tankers'. There followed a peculiar ceremony for the cameras at Tadcaster, where Hewitt's head brewer Barry Peacock pressed some buttons on a control panel to start the first brew for North Lincolnshire. 'Most impressive', he said, with perhaps some tongue in cheek.

An air of decline set in at Grimsby after brewing ceased, exacerbated by an enormous fire at the old Victoria Street maltings in January 1975, long abandoned by Hewitt's. As we've seen, it was the second destructive fire to ravage the building, and although it was listed, the damage was so severe that it had to be demolished. Surely the final blow, though, was the replacement of the Hewitt name in 1980, when all their former pubs were rebranded under the name William Stones, referring to a Sheffield brewery within the Bass empire. It was a name that meant little to drinkers missing a pint of Hewitt's HB bitter; Stones had never had much presence in Lincolnshire. Some effort was even made to cover up 'Hewitt Brothers' and 'HB' lettering on the surviving pubs. The Pasture Street buildings were quietly abandoned by Bass North Ltd in the early 1980s and left to the mercy of arsonists. They were finally demolished in 1985 and the brewery site is now occupied by the Great Grimsby Combined Court Centre, opened in 1988, a building whose design has some vague nods to the old brewery.

All is not lost. Bass slowly began to realise that the Hewitt name was still popular in the town, and in 1984 hitched themselves on the themed pub bandwagon to create Hewitt's Tavern in a pub opposite the brewery site, full of memorabilia from the old brewery, with a distinctive Hewitt's barrel logo fixed to the front wall. It's since returned to its old identity as the Duke of Wellington, a pub rebuilt by Hewitt's in 1940; the Hewitt's sign has unfortunately been relegated to the beer garden. The old cooperage in East Street, opposite the brewery site, was sensitively restored in 2020 after years of neglect, creating twenty one- and two-bedroomed flats. A large barrel jutting out from the top floor emphasises the building's past. And from time to time, Hewitt's beers are

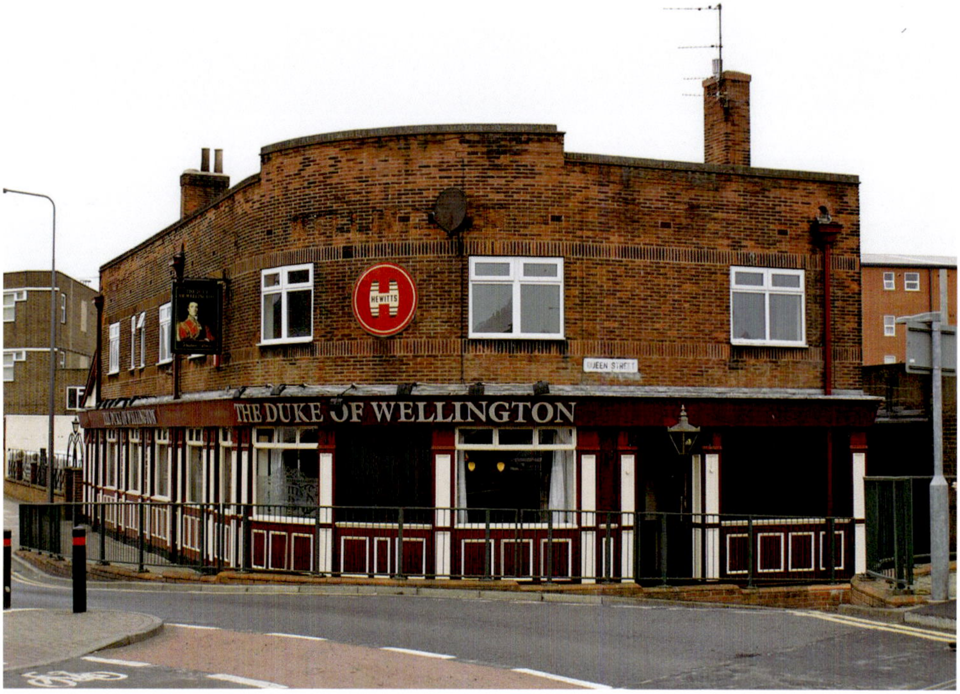

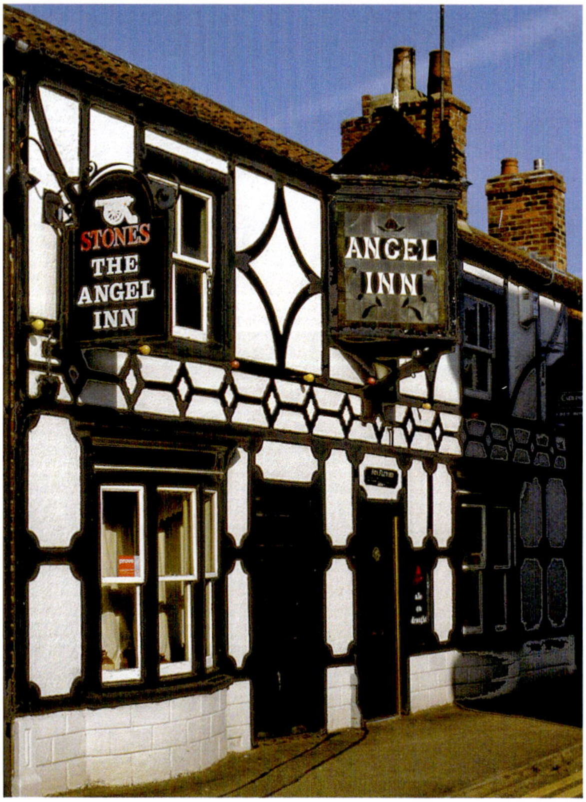

Above: The Duke of Wellington in Grimsby's Pasture Street, seen in 2011 complete with its red 'barrel H' Hewitt's sign, installed when it was briefly renamed Hewitt's Tavern. (John Law)

Left: Horncastle's Angel Inn, North Street, is now Montebello's Italian restaurant, but fortunately the coloured glass lantern sign survives, though not the less impressive 'Stones' version seen in this image from 1999. (John Law)

resurrected by enterprising craft brewers, the latest incarnation being Hewitt's 1874 Ale, brewed in Louth to an original Hewitt's recipe. But maybe the best outcome is that brewing is back in Grimsby with Docks Beers, a modern microbrewery in King Edward Street which opened in 2018, not far from Hewitt's old maltings. With a live music venue upstairs, and an eclectic range of beers, it's superficially quite different from Hewitt's until you realise that Docks Beers are firmly rooted in the Grimsby community just as much as Hewitt's once were.

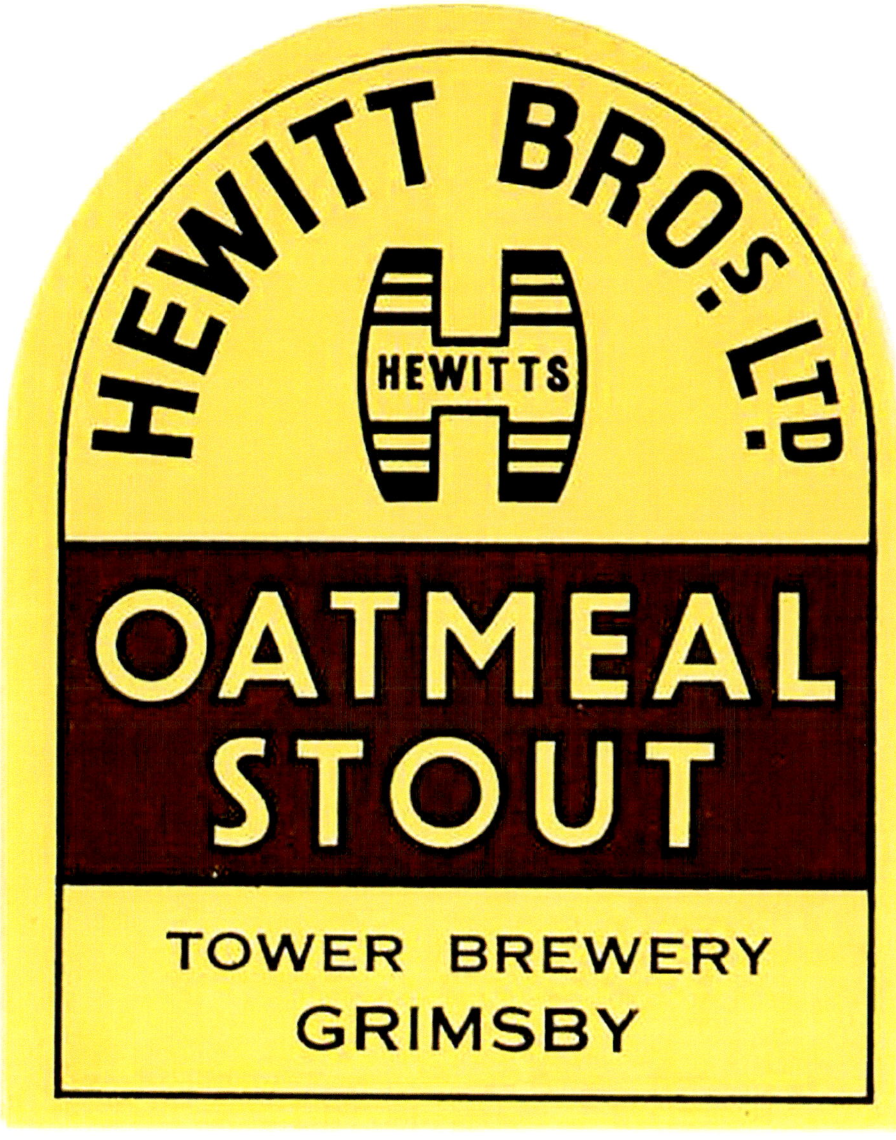

Hewitt's Oatmeal Stout label. (AC)

2

Melbourn's of Stamford

The year 2026 marks 200 years of brewing – albeit not continuously – at Lincolnshire's oldest brewery, All Saints, situated in the market town of Stamford. The brewery has passed through a wide range of owners, and although it's best associated with the firm of Melbourn Brothers, they only ran the operation for half of that time.

The founder of All Saints was William Brown Edwards, who'd decamped from a brewery in Water Street that would evolve into Hunt & Co., one of the town's other three major breweries. He built his new brewery on a prominent corner site in All Saints Street, beside the Great North Road as it wound through the town, right opposite the

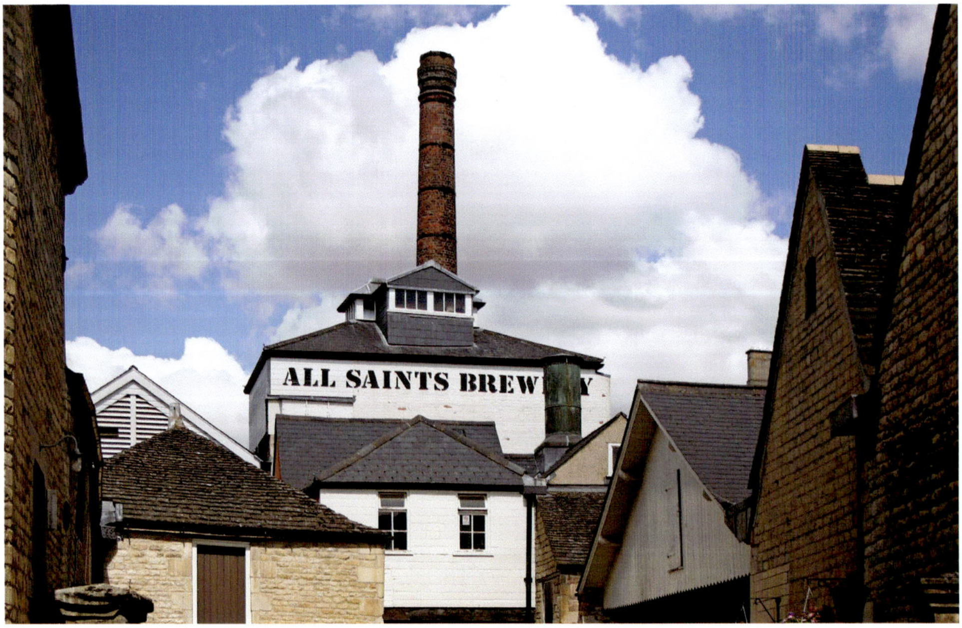

The brewery seen in 2007 from the yard, with the brewhouse and chimney dominating this view. The assorted buildings reflect growth and additions over the years: a perfect evocation of brewing history. (Richard Croft)

thirteenth-century All Saints' Church. Edwards' new brewery opened for business in January 1826. He was not short of money – his family had banking interests, despite a bankruptcy in 1814 – and within a few years he'd acquired a small estate of pubs, as well as constructing an elegant Italianate house for himself on Scotgate, just behind the brewery. Another brewery in North Street, Bourne, was acquired in 1845, and by the 1850s fifteen pubs were being supplied.

On Edwards' death in 1849, the business passed to his thirty-four-year-old son, also called William. All Saints was being described as a 'steam brewery' by 1852, indicating the installation of a steam engine to work the plant. Perhaps due to this investment, Edwards Jr did not enjoy his father's success and in 1857 declared himself bankrupt, several thousand pounds in debt. Some extremely murky dealings took place involving the theft of brewing equipment from the premises by William Young, who was almost certainly Edwards' brewer, and a court bailiff named Reeves. Both men were seen in the small hours whilst striding hastily towards the main line railway station at nearby Essendine, and were later arrested, on the strength of information in a letter from a member of Edwards' family. At the men's trial, Edwards gave some very evasive evidence, but admitted asking Young to 'look after' some goods for him, although refused to say whether any of them were the goods in question. He was probably fortunate not to have been charged as well; Young's barrister remarked pointedly how anxious Edwards seemed to have Young sent to prison. Perhaps his higher social status helped him. Though his days as a brewery owner were finished, Edwards' bankruptcy was soon discharged and he popped up a few years later working as a brewer with Lowe, Son & Cobbold, another of Stamford's brewing companies. Having served two months' imprisonment, William Young also returned to brewing in Stamford, and although it can't be proved, I suspect he resumed working for or with Edwards.

All Saints now entered some years of uncertainty. After the auction of its entire contents and stock in September 1857, a young man called Frederick Phillips, from Crowland, took control of the brewery, working in a loose partnership with the landlord of the Three Tuns pub in his home town. The whole venture lasted no more than eighteen months, however, and in August 1859 wine merchant Charles Walker took over from him. Walker proved no more durable, control passing to Elmer Brown two years later, who had been briefly in partnership with him. Brown, who had sold off a large farm at Uffington to join Walker, lasted six years before the money ran out and he was forced to declare bankruptcy in June 1867.

This unhappy procession of failed businessmen must have given All Saints a forlorn, neglected and perhaps rather ominous reputation, and it seems to have remained empty for a couple of years. One man not discouraged, though, was Herbert Melbourn, the son of a prosperous north Lincolnshire farmer from the village of Ingham. He had a good knowledge of brewing, having learned the trade over a few years in Germany, when he'd also gained an understanding of wine and spirits. There's a good chance he was looking around for a suitable business, and must have been attracted by the potential of All Saints brewery and its town centre site, despite its likely run-down appearance and the many unsuitable proprietors who had gone before. He took over All Saints in early 1870, at the age of twenty-two, and would manage to do what none of his predecessors could: create a lasting and profitable business there. He soon had brewing back up and running, and quickly began to build an estate of pubs. A short-lived partnership with Herbert's brother George ended in 1874, but even though Herbert ran the show on his own for many years, it was always known as Melbourn Bros. Probably due to the family's connections in northern Lincolnshire, a number of pubs in the Lincoln area were owned. As they were 50 miles distant from Stamford, a Lincoln warehouse at Brayford Pool was maintained to serve them, and to handle George Melbourn's farming supplies trade, which was operated until 1924.

Herbert married Lavinia (Fanny) Bullivant in 1881, and thereafter led the lifestyle of a prosperous businessman, travelling widely and living in fashionable seaside resorts. He ensured, however, that he had reliable local management. An astute decision in the early days had been to hire Henry Waite from Tonbridge in Kent, who joined as a very young man, rising to be head brewer and manager. Herbert also invested in the brewery; having built a new brewhouse in 1870, he added a hop back and malt tun over the next few years. In 1888 Herbert's younger brother Stanley joined the business following the sale of significant farming assets at Ingham, which brough further investment capital into the business. By the end of the nineteenth century the brewery had been modernised with all new plant. A new coal-fired boiler in 1902 completed the programme of investment and the brewery continued to use its original equipment largely unaltered for the following seventy years. Indeed, much of it remains in place to this day.

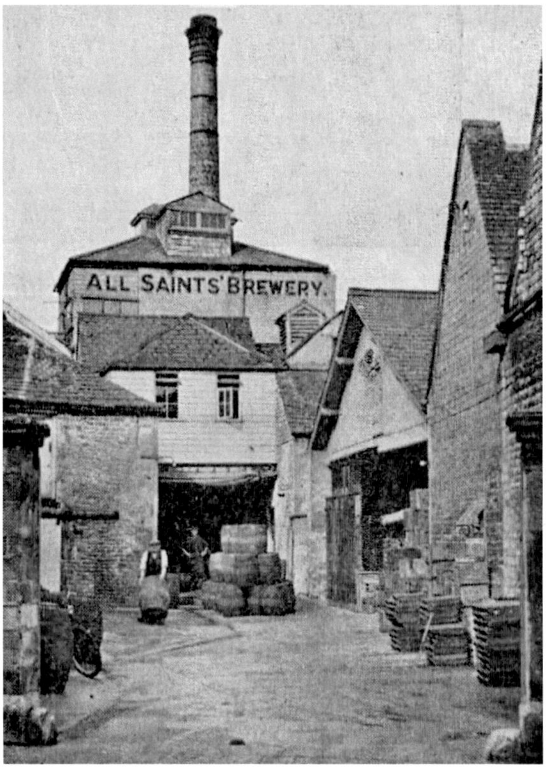

The brewery yard seen in the early 1950s from a Melbourn's advertising brochure, from a similar viewpoint to the image on p. 26 but with much more activity going on as barrels are moved down the courtyard. (Brewery History Society)

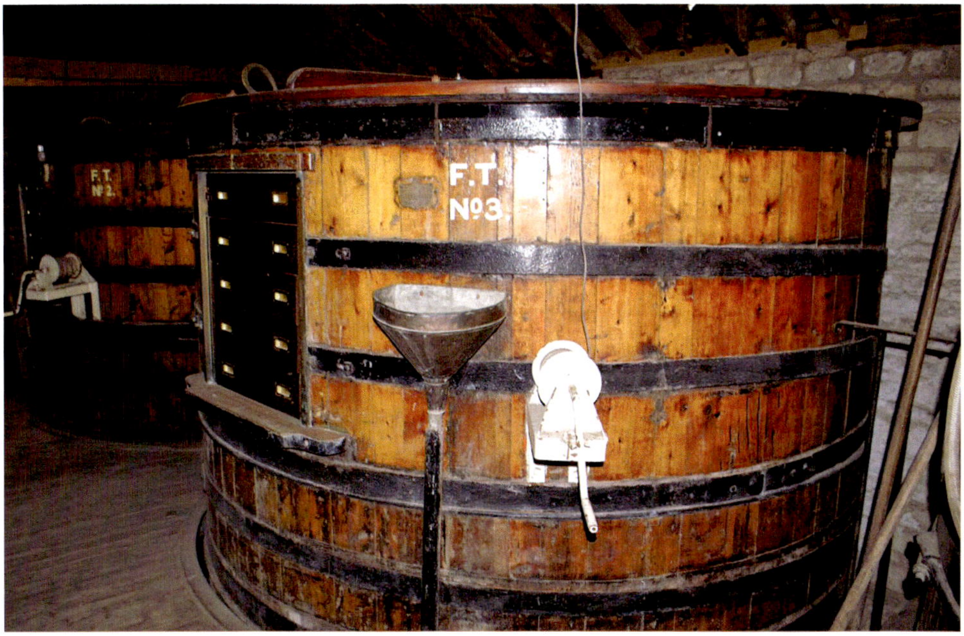

Two of the three white cedar fermenting vessels, numbered F.T. No. 2 (left) and F.T. No. 3 (right) seen in 2009. The trio were supplied in 1873 by Shuters and Chippendales, a cooperage firm from Stratford, London. (John Law)

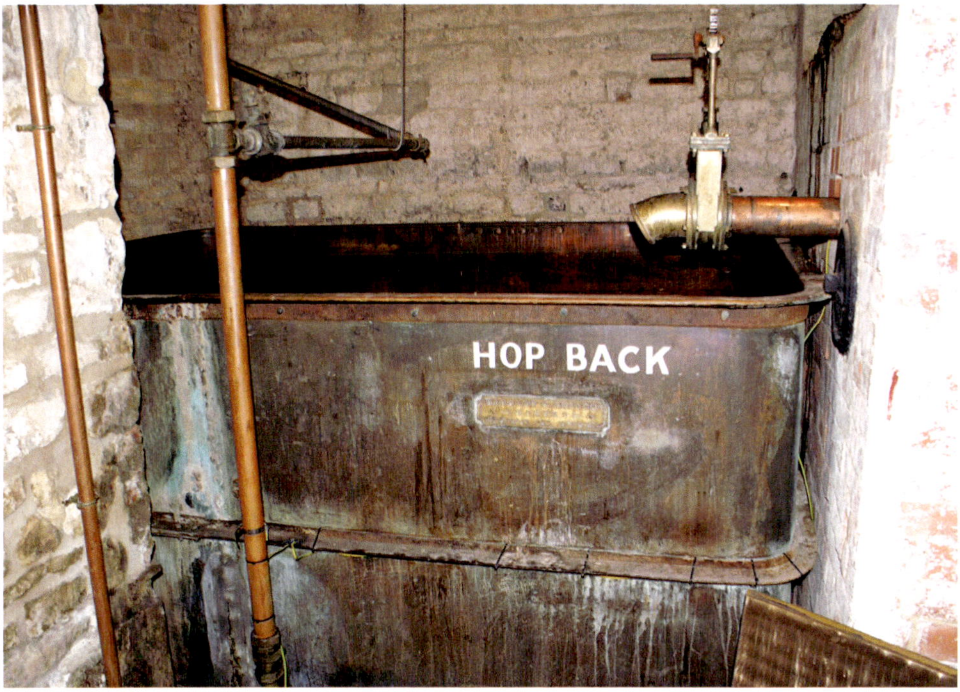

The hop back was of a similar vintage to the fermenting vessels; it too was still in daily use up to the 1970s. (John Law)

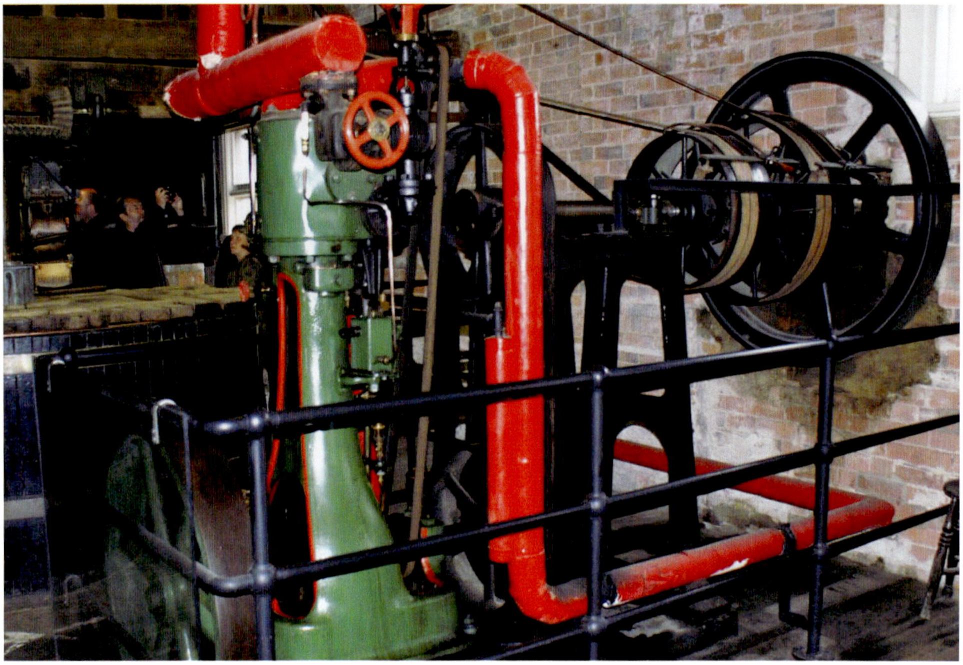

The steam engine in All Saints brewery, believed to have been built by the engineering firm Marshall's of Gainsborough. (John Law)

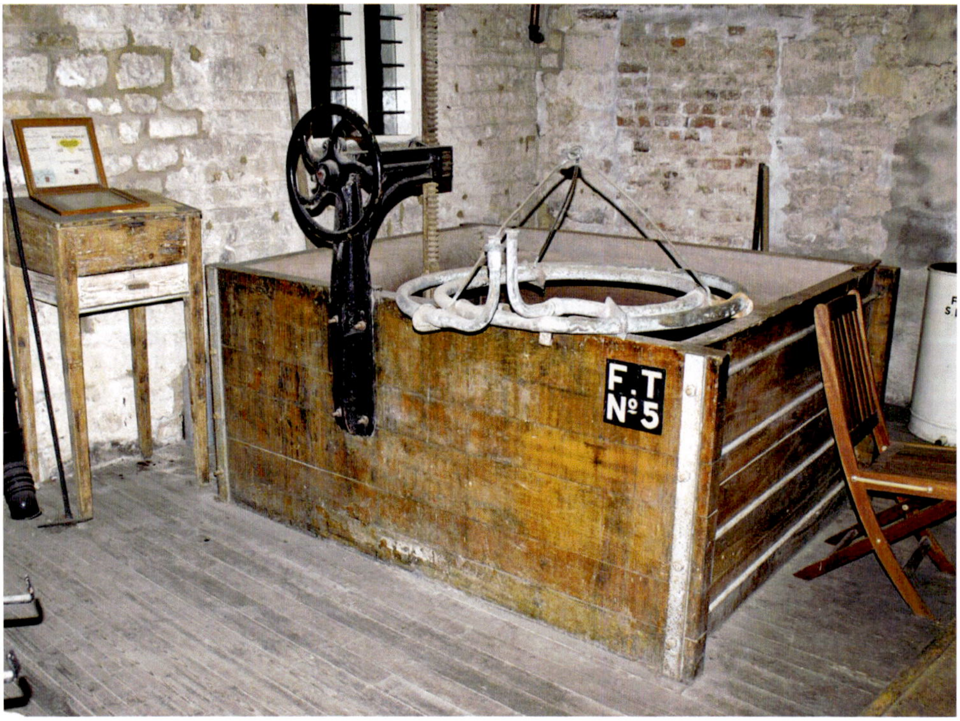

Square fermenting tank No. 5 at the brewery in 2009. (John Law)

The tied estate had dwindled over the years, declining to around twenty by the end of the First World War as licence renewals were refused for some of their less desirable pubs. A programme of gradual expansion and improvement of the estate was therefore started. The Royal Oak at Heckington was completely rebuilt in 1931 on a prominent corner site, and four pubs in South Lincolnshire were also bought that year to add to a couple acquired in the late 1920s. This gradually brought Melbourn's pubs back up towards thirty, enabling them to compete a little better with rival breweries.

The years were starting to catch up with the Melbourn brothers and their key staff. Stanley continued living in Lincolnshire and oversaw the Lincoln operations from his home at South Carlton until his retirement in 1925. Head brewer Henry Waite continued working until his death in 1923 at the age of seventy-three, when local man Walter Walmsley took over. Herbert died at Bournemouth in January 1927 at the advanced aged of eighty-seven, followed by Stanley in 1935. The business was restructured into a limited company, ownership remaining within the Melbourn family, however. Day-to-day management continued in the hands of a local manager, initially Walter Walmsley, but in 1938 Edward Godwin from Swindon was appointed, following the sale of his family's brewery there.

Herbert's children Max and Christine both lived in southern England, visiting Stamford quite rarely and taking little active part in the business. Max's talents lay outside brewing; he was a champion budgerigar breeder and judge, and by way

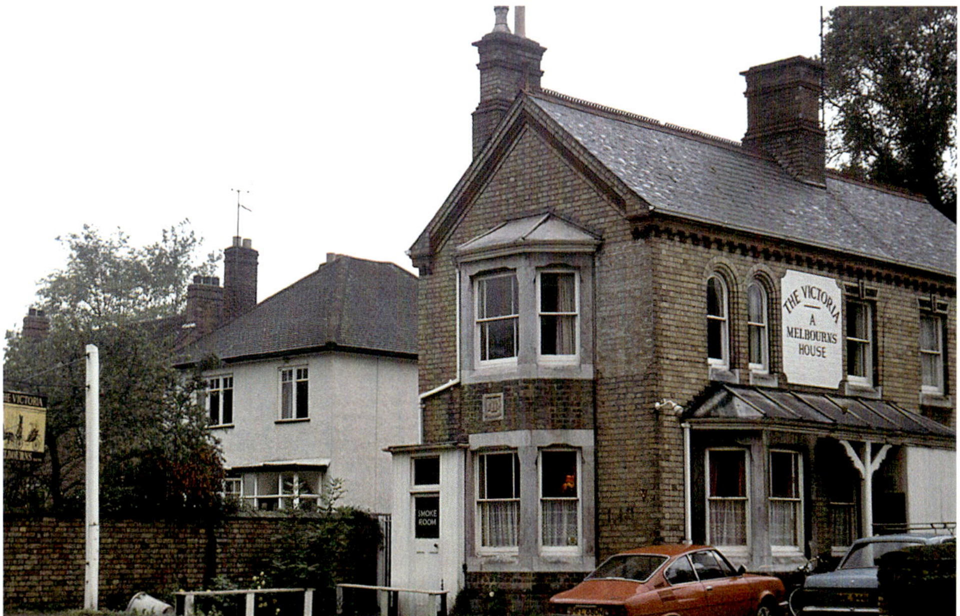

The Victoria pub in Stamford, seen in 1976 proudly declaring its ownership, standing on Ryhall Road on the way out of Stamford. It was built by Melbourn's in 1887 to replace the Parting Pot inn and patriotically named after the Queen. Now a private house, it still carries the remnants of the white painted sign. A small end terrace house to its right is named Parting Pot Cottage. (John Law)

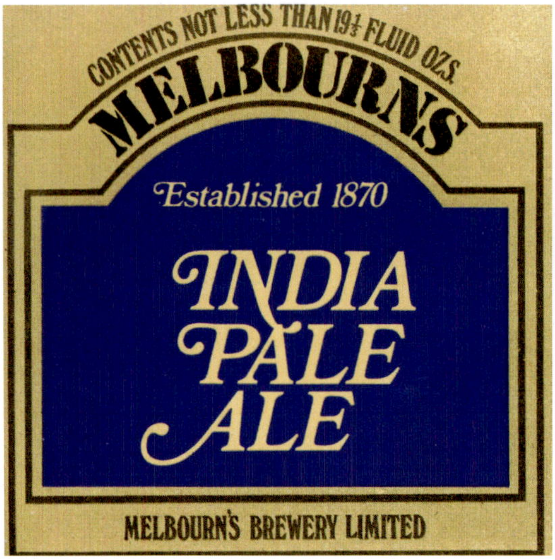

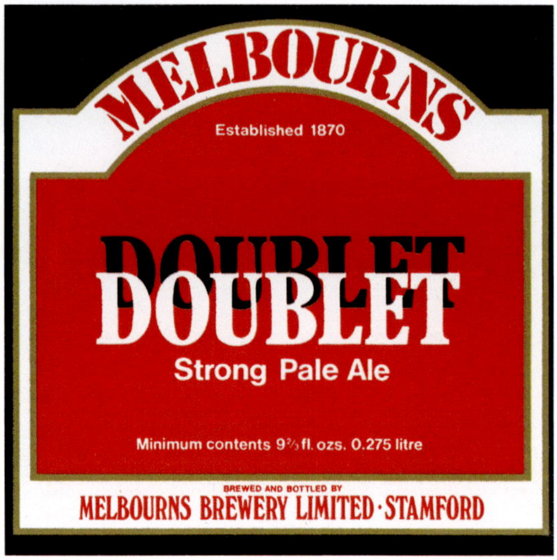

Above left: Melbourn's IPA label. (AC)

Above right: Melbourn's Doublet label. Both of these labels date from around Melbourn's centenary in 1970. (AC)

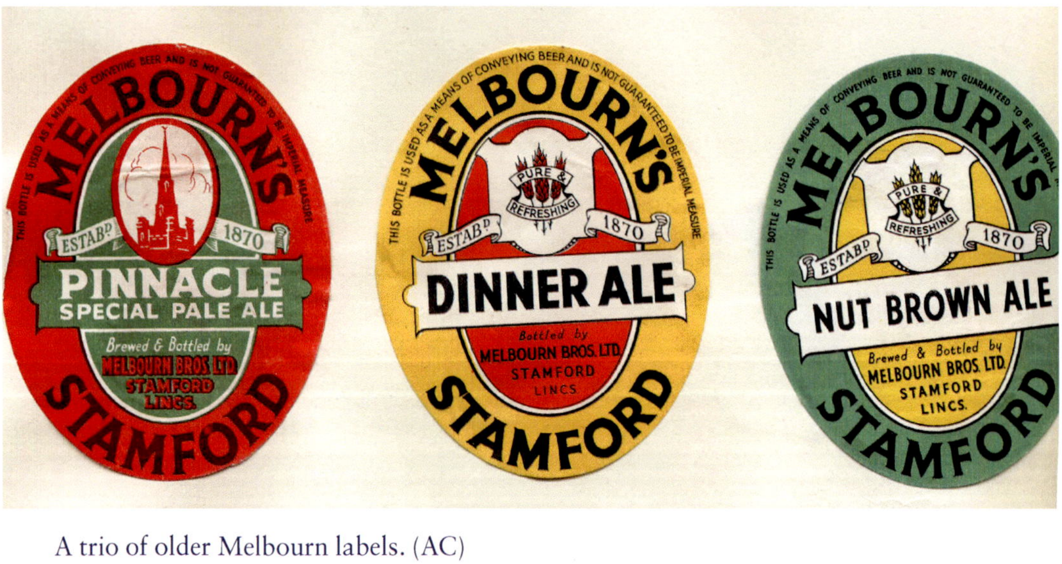

A trio of older Melbourn labels. (AC)

of contrast, was also a Merchant Navy officer, volunteering for military service in the Second World War, when he commanded a minesweeper. The family remained largely hands-off owners, trusting key employees to get on with running the brewery. It was a policy that worked well: production of bottled beer began to increase and in 1950, Melbourn's bottled pale ale won a national competition, although their draught IPA still remained popular. Under the stewardship of head brewer Dick Walmsley, who'd taken over from his father, Walter, in 1939, a new premium bottled ale, Pinnacle, was introduced early in 1961, followed that summer by a special strong ale, Charter, to celebrate 500 years of Stamford's corporation. Then their Nut Brown bottled ale secured first prize in the 1968 Brewers' Exhibition. But the tied estate remained around thirty, mainly comprised of smallish village and town pubs in attractive locations along the limestone belt in western Lincolnshire and northern Northants. These were numbers far below the interest of any corporate raider, so the increasingly ancient, and still steam-driven, All Saints Brewery continued to brew twice a week, producing high-quality, low-volume beers that bucked the national tide of bland keg beer. In 1970, the company celebrated its centenary, sharing the event with their tenants and twenty-two brewery staff, six of them with more than twenty-five years' service.

This amiable existence couldn't continue forever, and just one year later the family announced that the business was for sale. Within two months, a deal had been tied up, but when the buyer was announced in July 1971, it was a surprise: not a national firm, but two members of the local aristocracy, the Earl of Gainsborough and his son-in-law, the Earl of Liverpool. From the family seat at nearby Exton Park in Rutland – family motto: 'All well, or nothing' – the two men managed a large estate of land and property. The pub trade was not a complete unknown for the Earl of Gainsborough, who had previously owned a handful of pubs as part of the Exton estate portfolio, which were leased to breweries, but the pair had never overseen pubs or a brewery. And the purchase created a fairly daunting prospect for hopeful job applicants, who had to send applications to the Earl of Liverpool.

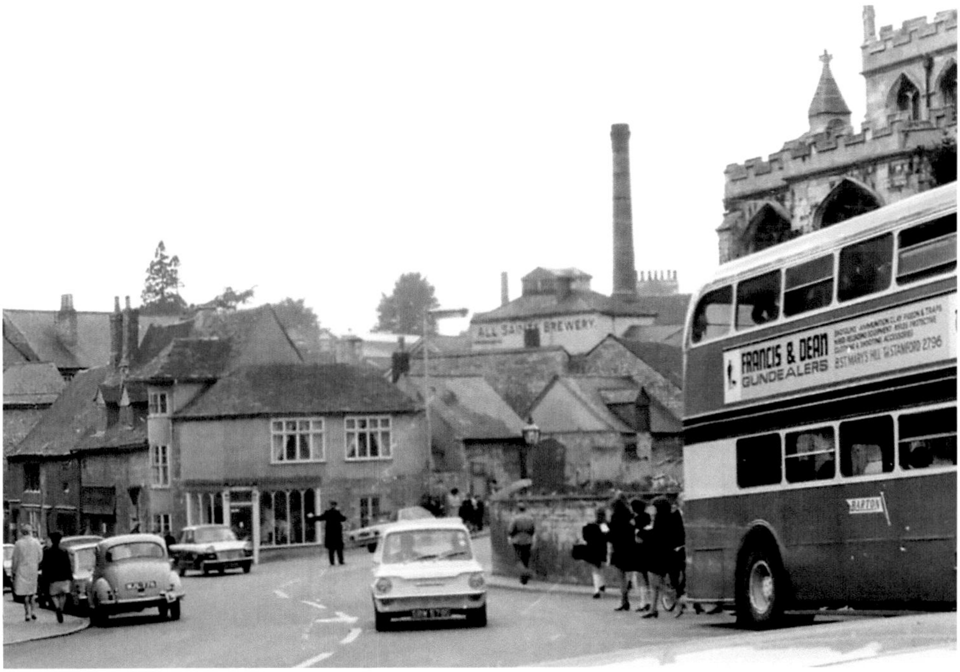

Melbourn's All Saints brewery in its town centre context, seen in 1972 with a policeman on point duty ensuring traffic keeps flowing on the old Great North Road. The brewery chimney sails upwards to the left of the Barton bus whilst All Saints' Church keeps a serene watch on proceedings. (Brewery History Society)

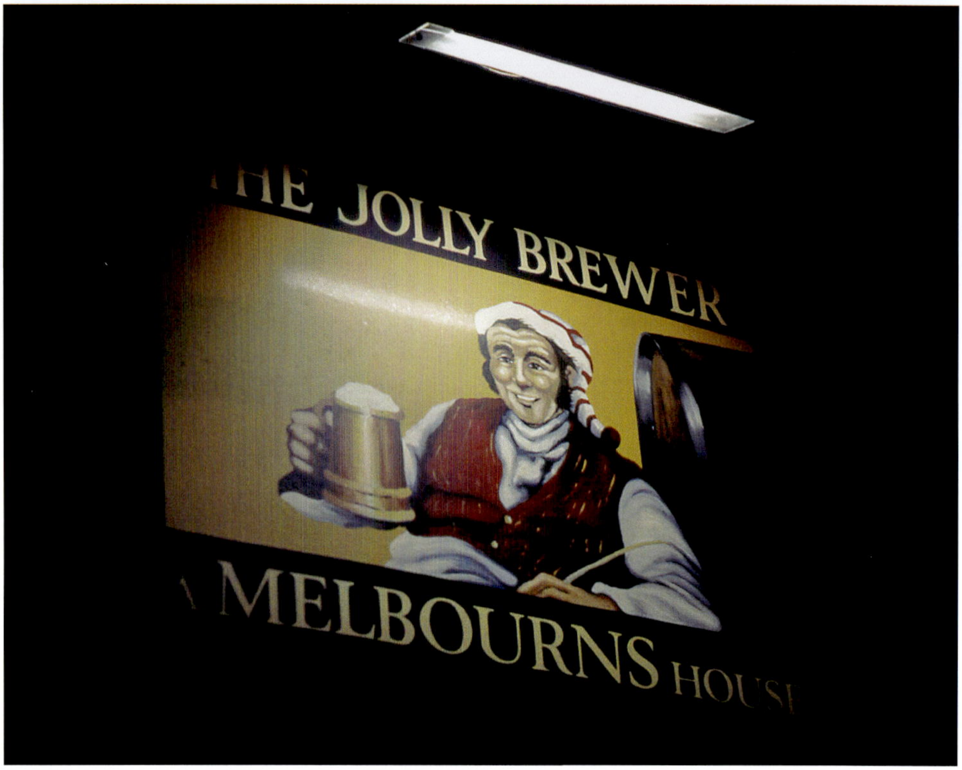

A comforting night-time image of the Jolly Brewer pub sign in Foundry Road, Stamford, still displaying its Melbourn ownership in 1988. This pub remains open. (John Law)

Despite this promising news, the economic problems of the 1970s, especially soaring inflation, were beginning to bite. What really hampered the new owners, though, was the discovery of serious and expensive issues with the brewery's boiler which dated back to 1902. All of these concerns pushed them towards selling the business. In April 1974, the purchaser was revealed to be Samuel Smith's of Tadcaster, Yorkshire, a brewery founded in 1757 which projected a strongly traditional image and was expanding into the London free trade market. However, Smiths had only bought the brewery, not the pub estate; the deal was that they would supply Melbourn's houses with beer, with Melbourn's becoming what we'd now call a pubco – a business that owns pubs but doesn't supply them. Further unexpected news came with Smith's decision to re-introduce a traditional brewers' dray cart, a method Melbourn's had been using as late as 1933. A dray and two Shire horses, named Mild and Bitter, were accordingly based at All Saints, although their deployment was limited to Stamford town centre deliveries. This rejection of modern methods made a great spectacle for tourists although perhaps less so for the town's motorists, increasingly beleaguered by traffic problems. Brewing ended at Stamford in June 1974, and although the site was then used as a distribution depot by Smith's, most of the former brewery staff were made redundant the following year.

The surprises hadn't finished, though. The antiquated old brewery with its rambling rooms, wooden mash tuns and polished equipment had great charm and attraction. Lord Liverpool had previously spoken of his hope that it might become a museum, and that's what eventually happened under Smith's ownership. In July 1978 the Stamford Brewery Museum opened, welcoming visitors five days a week during the summer months. However, by the mid-1990s the museum had closed in favour of a new brewing initiative: fruit beer. Smith's managed to restore the old brewing equipment, coaxing it back into life to produce organic cherry, strawberry, apricot and raspberry bottled beers, which are all still quite widely available. Though the fate of the museum is unfortunate, it's heartening that All Saints still occasionally produces beer, especially as the brewery's mash tun of 1876 is now the oldest still used in the UK.

Melbourn's carried on for another thirteen years, picking up a few more pubs along the way, including two from Ruddles of Oakham when they sold off nearly all their tied houses in 1978. The booming economy of the late 1980s fostered a renewed climate of corporate takeovers, as investors with little connection to brewing, and sometimes even less understanding of it, sought to diversify their operations. That was the backdrop to the end of Melbourn's in 1987, when their Lordships sold the remaining business, including thirty-two pubs, to Camerons of Hartlepool, an acquisitive operation which had been City owned since 1975 and was by then in the hands of the Barclay brothers (Camerons are now independent again). A bewildering procession of further takeovers followed; suffice to say that in 1991 many of the former Melbourn pubs were spun off to one of the new mega-pubcos, Pubmaster. A pleasingly high number remain open in 2025.

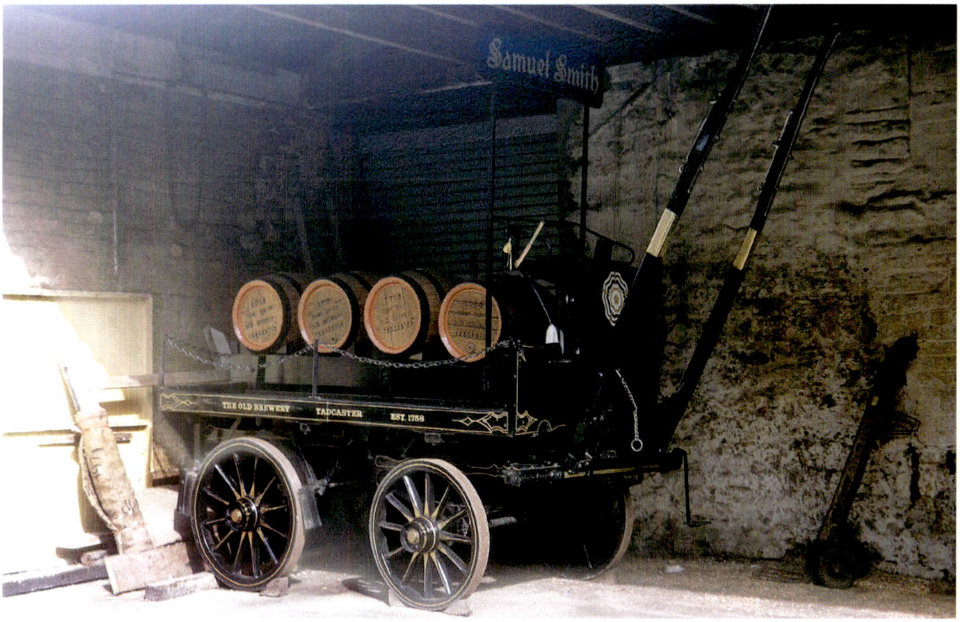

Samuel Smith's brewery dray on display when the Stamford Brewery Museum was still open. (John Law)

3

Mowbray's of Grantham

Mowbray's were one of the oldest and longest-lived breweries in Lincolnshire. However, less than a decade after it was founded, the Mowbray family had lost all control of the business, in somewhat controversial circumstances. The story of how a group of local businessmen effectively seized control and ran the firm for over a century, eventually owning over 200 pubs, is worth a closer look.

The business was started by Samuel Hall Mowbray, born into a Fenland family at Butterwick, near Boston, in 1796. He knew the brewing trade, having set himself up as a common brewer at the Green Dragon inn in Boston. He moved to Grantham at the age of twenty-seven, where he ran the Granby inn in the marketplace for a couple of years until he took a lease on some land in Commercial Road. There he started building a brewery, which opened for business in September 1828. However, a dispute soon arose over the land he'd built the brewery on, and Mowbray needed financial help to survive. He threw himself into the hands of two local businessmen, Frederick Robertson and John Burbidge. According to the Mowbray family history, a pithy but not always accurate document, a copy of which resides in Lincolnshire Archives, Robertson and Burbidge initially loaned Samuel Mowbray sufficient funds to continue trading, but when by 1836 he could not repay them, they threatened to 'ruin' him unless a three-way partnership was formed.

With no other clear options available, Samuel Mowbray signed himself into this joint arrangement. For several years afterwards, he divided his time between Grantham and the Fenland village of Gedney, where he set up a much smaller brewery with a man called Septibo English. Perhaps he had hopes of a new start there; certainly, his involvement with the Grantham brewery seems to have ended. In later life, Samuel's health was poor; he suffered from respiratory problems and by 1851 he was living in Oldham. Moving to the heavily industrialised north-west was a strange choice for a man in his state of health, but Samuel had set up the Waterloo Brewery there, another small-scale venture. He later retired to Lower Broughton, a Manchester suburb, where he died from chronic bronchitis in 1864, aged sixty-seven. It was a sad end, miles from the prosperous firm he had created, and the family history talks bitterly of the 'plunder' of the family business, with Samuel's sons and daughters being barred from the brewery. Somebody, it says, had robbed the Mowbray's 'from right to left' – not Robertson nor Burbidge, but a man called William Grasby, the executor of Samuel's will.

Grasby had several professions down the years – commercial traveller, draper, shipping agent, wine merchant – and was briefly on the payroll at Mowbray's. When the brewery became a limited company in 1888, he was one of the shareholders, but it seems unlikely he profited to the extent that the Mowbray family believed, nor did he hold a quarter of the shares as they thought; his last home was a small, terraced house in the Liverpool district of Toxteth. Grasby does seem an improbable villain, if there ever really was one; like many other early brewers, Samuel Mowbray may just have been very good at making beer but very poor at running a business.

In the meantime, the brewery was in the doldrums. It wasn't particularly well run, with a number of safety issues cropping up, and was often outbid by competitors when local pubs came on to the market. Things only started to improve when John Burbidge's grandson Arthur Hutchinson was appointed managing partner in 1882, aged only twenty-six. His first key decision was to branch out into fizzy drink production. Flavoured non-alcoholic tonics were becoming increasingly popular, partly due to the growth of the temperance movement, and this new line meant that Mowbray's no longer had to buy in carbonated waters. The incorporation of the firm

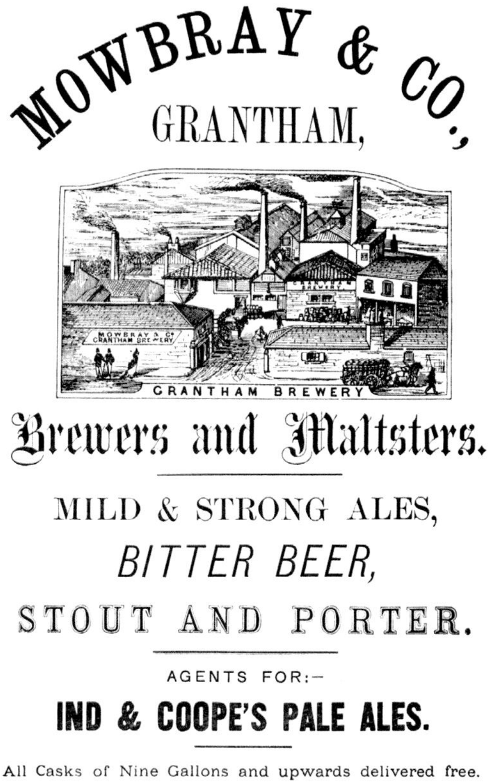

An advertisement for Mowbray's dating from the late nineteenth century. The drawing is a relatively accurate representation, unlike the imaginative depictions that often adorned similar publicity pieces. The three tall chimneys in the centre of the image were apparently named after Samuel Mowbray's daughters, Louisa, Mary and Priscilla. (AC collection)

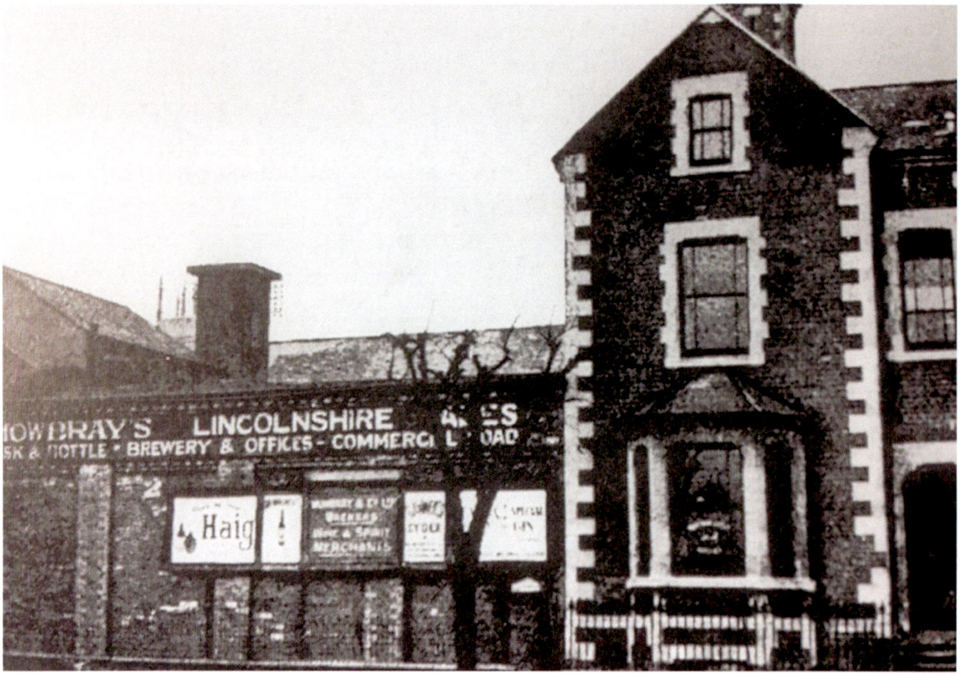

Photographs of Mowbray's brewery are rare, especially the London Road site, which was mostly demolished in the early 1960s. It's seen here in a grainy photograph possibly from the interwar period. The chimney of the former Redhead premises can be seen to the left and also in the photograph below. (Grantham Civic Society)

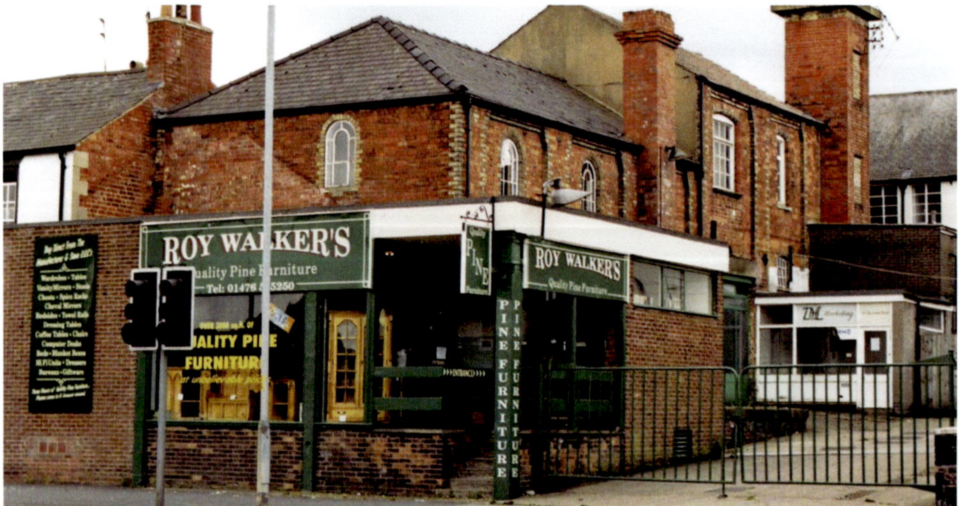

The former Redhead brewery stood at the southern end of Mowbray's main site on London Road, with the Barley Mow pub just visible to the left in this view. After complete closure, most of the site was demolished but the old Redhead buildings were used by a furniture retailer for many years. Even this has now gone, to be replaced by a small retail park. (Allan Senior)

as a limited company freed up capital for expansion, so when the Spittlegate Brewery in London Road, Grantham, came on to the market following a spectacular falling out amongst the Redhead family, it was quickly snapped up by Mowbray's. Their premises were very close to Mowbray's existing site on Commercial Road, and by piecemeal development the two sites came to work as one brewery, albeit a sprawling and not especially efficient one. As a result of the Redhead purchase, their brewer Arthur Rear joined the firm, and was still there when he died thirty-five years later.

Under Arthur Hutchinson's leadership, the new company quickly became very profitable, paying 10 per cent dividends on ordinary shares by the end of its third year, and almost double that within ten years. That impressive performance drove a shopping spree of pubs, mainly fairly basic ones in and around Grantham, but for the right price they would look further afield, splashing out several thousands on the Nag's Head (now the Ivy) in Sleaford and the Queen's Head in Boston (now closed). As the firm expanded, so did brewing capacity and staff; by 1891, the payroll had swelled to around forty. Long-serving staff were a feature of Mowbray's, like other similar firms. Herbert Wheeler had arrived from Sergeant's of Brigg in 1888; he rose to be company secretary, and his son Denis also joined the firm, becoming head brewer. They served a combined total of over sixty years at Mowbray's.

We move forward a few years to September 1905. Arthur Hutchinson is standing on the floor of Lincoln Corn Exchange, jostling for space with some 600 other hopeful

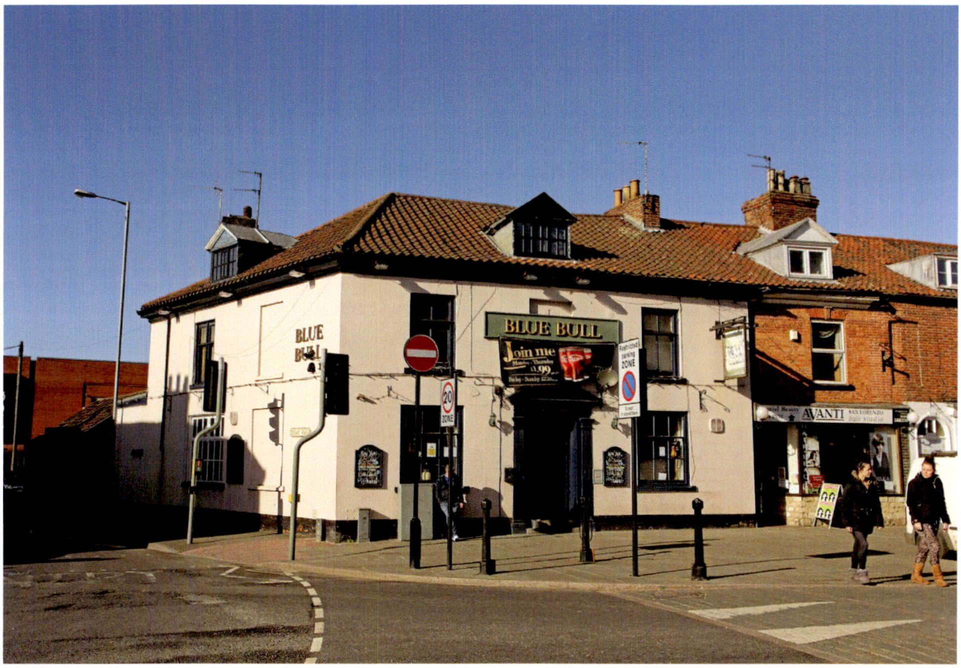

Grantham has several pubs with the word 'blue' in their name, which harks back to political rivalries some 200 years ago. Blue pubs were frequented by Whig supporters, of which one was the Blue Bull on Westgate, seen here in 2014. Mowbray's ran ten 'Blue' pubs across the town. (Richard Croft)

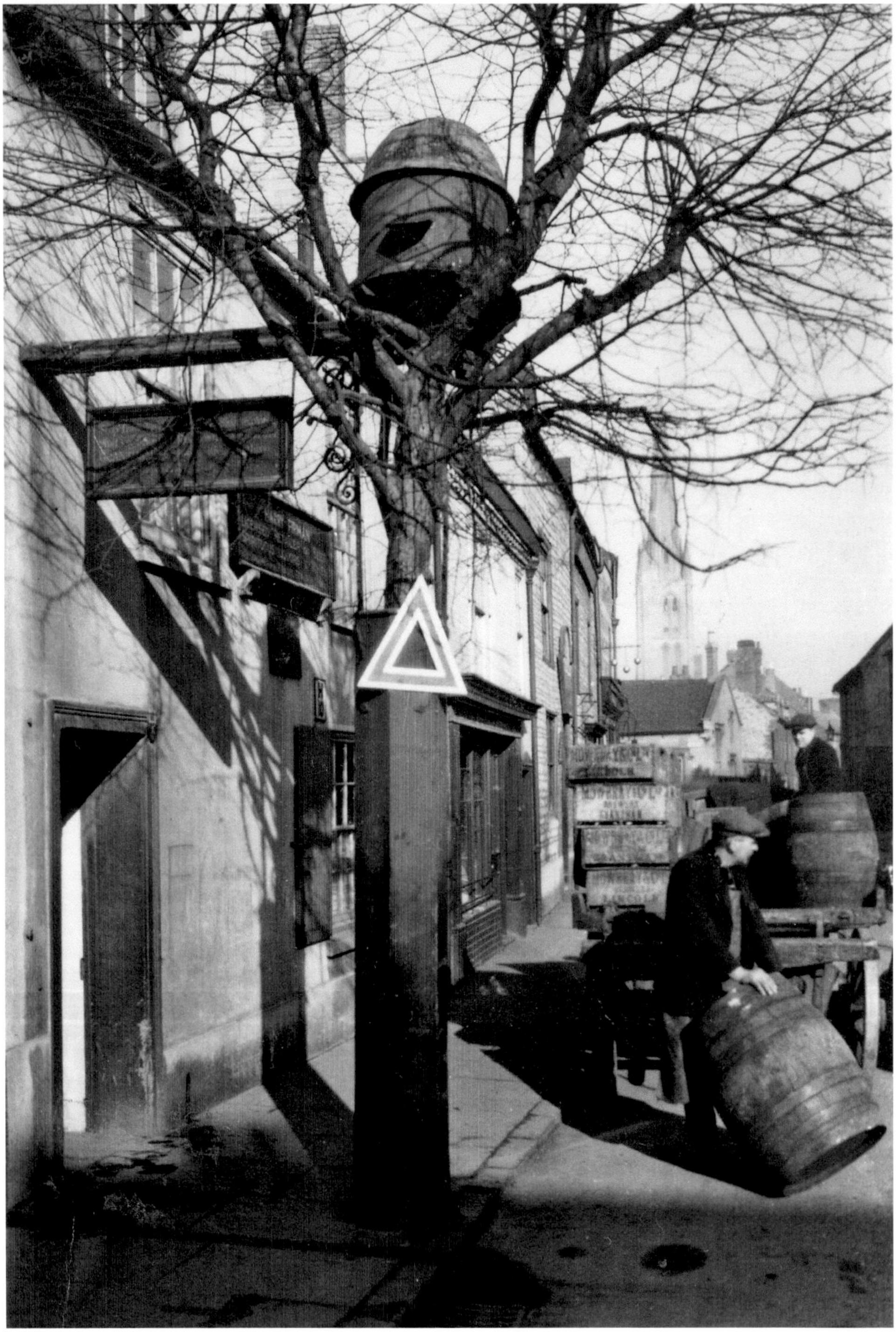

One of the most unusual pub signs ever must be this 'living' example at the Beehive in Grantham – an actual beehive. Mowbray's bought the house in 1918; both pub and sign are still in good health. (AC collection)

bidders and trying to blend into the background. It's the grand auction of Dawber's Brewery, a force in the city for nearly eighty years. With its fifty-two pubs in and around Lincoln, and a fair-sized brewery and malt kiln on Carholme Road, producing around 7,000 barrels of beer annually, it's proving a popular draw. Arthur has already had a good look round the brewery and many of the pubs with a valuer called Mortimer Jones. They like what they see: a successful bid will extend Mowbray's pub estate into the city and north of the county, so Arthur has taken the precaution of having Jones bid for him – he doesn't want to make his interest plain. The bidding starts. Others are being careful too, so the first bid, of £40,000, comes from Alexander Trotter, a local solicitor and the deputy coroner; he's obviously acting for someone else too. More bids, and the price rises by £5,000 each time. Now Jones is fighting it out with William Rasdall, the owner of the Saracens Head Hotel in Lincoln High Street; it's known that he is bidding on behalf of a syndicate. Jones bids again; the price creeps up. Rasdall bids, and Jones again. The auctioneer waves his gavel at Rasdall. 'Come on! Your shareholders will back you up! Your opponent is nearly done,' he claims. The gavel's about to come down, but Rasdall bids another £25. Jones outbids him again, and with a final cry of 'I've done!' Rasdall backs down.

The news that Mowbray's had purchased Dawber's was said to be well received by the latter's landlords; the other main bidders may have been Hewitt's of Grimsby and the Market Rasen Brewery. Mowbray's had paid £66,475, just above the reserve, or somewhere around £32 million in modern prices, giving them a combined tied estate of over 100 pubs. Brewing ceased at Lincoln almost immediately – that was a condition of the sale – and the Carholme Road premises were sold for £4,350, recouping some of the purchase price. As a counter to this good news, however, Mowbray's felt the blunt end of new legislation aimed to reduce the number of pubs. The 1904 Licensing Act gave local magistrates much wider powers to refuse pub licences when they came up for renewal if it could be proved the pub was 'unnecessary'. What that meant in practice and how to prove it tied up the police, landlords, brewers, temperance enthusiasts and sometimes customers for hours. There was little or no transparency around magistrates' decision making in such cases, but generally relatively small pubs with other licensed premises close by were targeted, especially if the landlord had a poor compliance record. 'Closed under compensation' – a limited form of parachute payment which paid out more to the owner than the landlord – became the epitaph for many luckless boozers. Mowbray's themselves lost around nineteen pubs between 1906 and 1939, just under 10 per cent of their estate, and then had to wrangle like everyone else with the judiciary about an appropriate compensation figure. The first pub they lost in this way was the Railway Tavern at Ancaster, a small beerhouse (that is, not licensed to sell wine and spirits). There was good evidence that the pub was well run with no complaints ever received about it; it sold almost three barrels of beer a week. Mowbray's beers had an excellent reputation in the area and were popular with the local quarry workers. Despite all this, the magistrates decided to close the pub, throwing the landlord out of work. As he'd only been there for six months, he got £60 in time-apportioned compensation compared to Mowbray's £1,128.

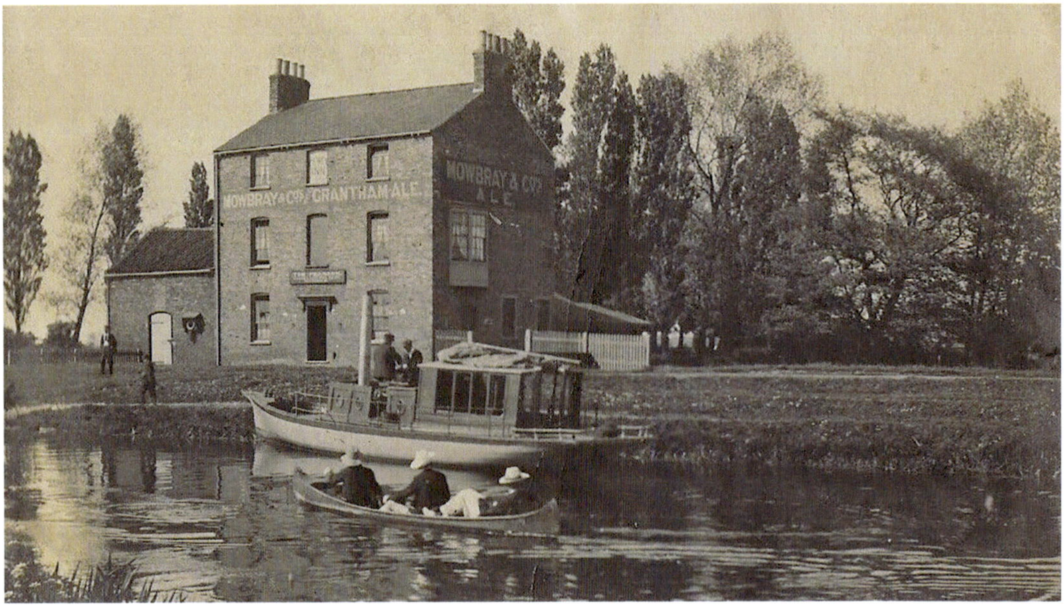

A postcard view of the Pyewipe Inn on the Fossdyke navigation west of Lincoln, probably not long after Mowbray's bought the pub and some time before it was extended. The prominent advertising was designed for maximum visibility. (AC collection)

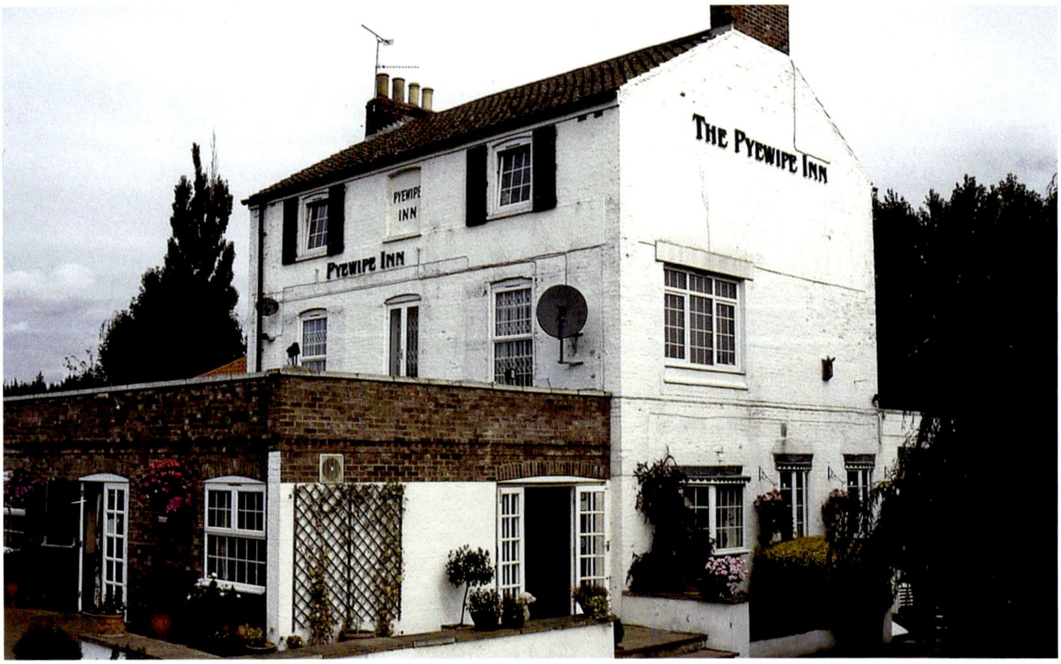

The Pyewipe Inn was acquired, along with fify-one other pubs, when Mowbray's bought Dawber's Lincoln brewery in 1905. It's seen here in 2011, with a modern ground-floor extension. (John Law)

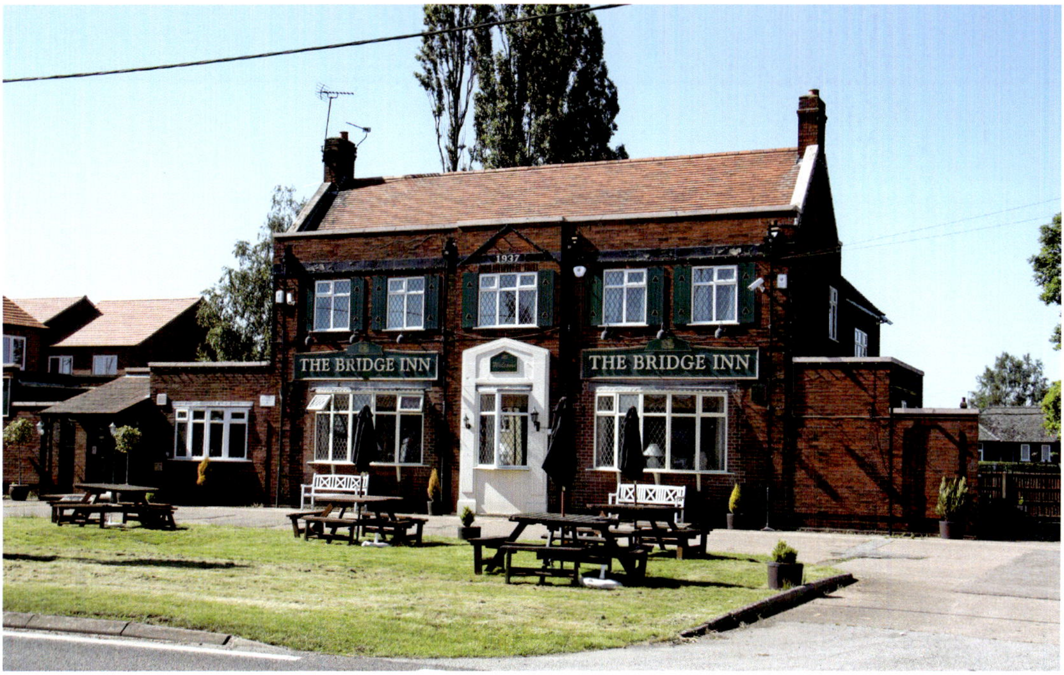

One of the pubs rebuilt by Mowbray's in the 1930s was the Bridge Inn at Saxilby, which reopened in 1937 as the building itself confirms just above the middle upper-floor window, seen in this 2010 photograph. The pub sadly closed shortly afterwards. (Richard Croft)

Despite these setbacks, Mowbray's continued to make good money. By the end of the First World War they owned or supplied about 200 pubs in and around Grantham and Lincoln. In 1926, they stepped in to buy the Stamford brewery of G. & H. R. Hunt and their sixty-five pubs in south-west Lincolnshire, Rutland and north Northamptonshire. To fend off other interested brewers, they ended up paying £140,000, more than many thought the firm was worth. They had to issue £200,000 in new preference shares to finance the deal, giving them room for further expansion. A Grantham wine merchant, Johnson Basker, was taken over in 1933 and in 1935 the wine and spirit business of James Thorpe of Uppingham, Rutland, was acquired. There was money for new pubs too. Their first, opened in December 1927, was the Earlesfield Hotel (now the Priory) on Dysart Road, Grantham. It was designed to have an upmarket and stylish appearance, a tactic that played well with licensing magistrates. Eight more new or rebuilt pubs followed in the 1930s: three in Lincoln and one each in Grantham, Potterhanworth, Stamford, Peterborough and Saxilby. All but one of them have survived to this day, although some are in alternative use.

Arthur Hutchinson retired as managing director in 1925, although he remained on the board until he died five years later aged seventy-five, unusually leaving over £1,000 to his employees. The board grumbled about the problems of finding his successor, with a number of short-lived appointments that didn't really work out. One that did was Hutchinson's daughter, Doris, who joined the board in 1929 and served as company

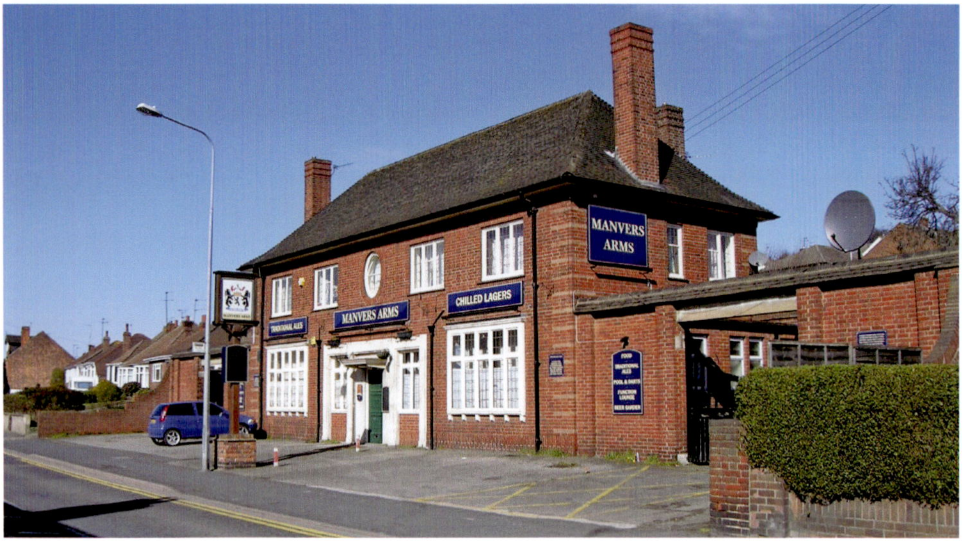

Most of Mowbray's new or rebuilt pubs of the interwar period were bulky, four-square affairs, and a typical example was the Manvers Arms in Monks Road, Lincoln, opened in 1934 to replace a city centre pub of the same name – it was first called the New Manvers Arms. It closed not long after this 2007 view and has been replaced by housing. (Brian Mossemenear)

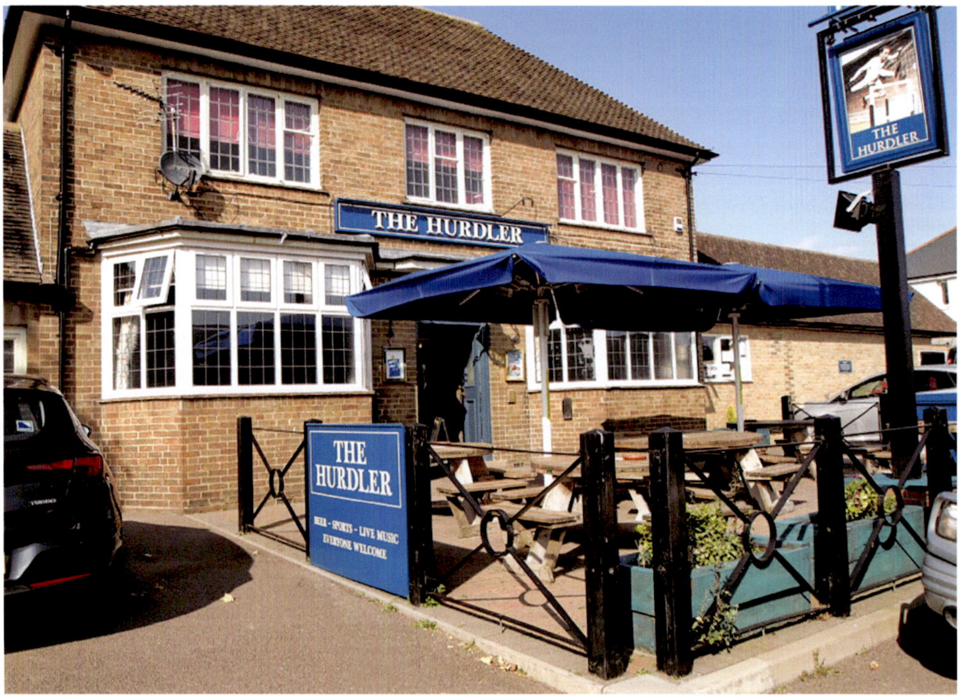

The Hurdler in New Cross Road, Stamford, an estate pub built by Mowbray's in 1937, named to honour the athletic skill of the 6th Lord Burghley, who won a gold medal at the 1928 Olympics. The pub remains popular. (Mark Shirley)

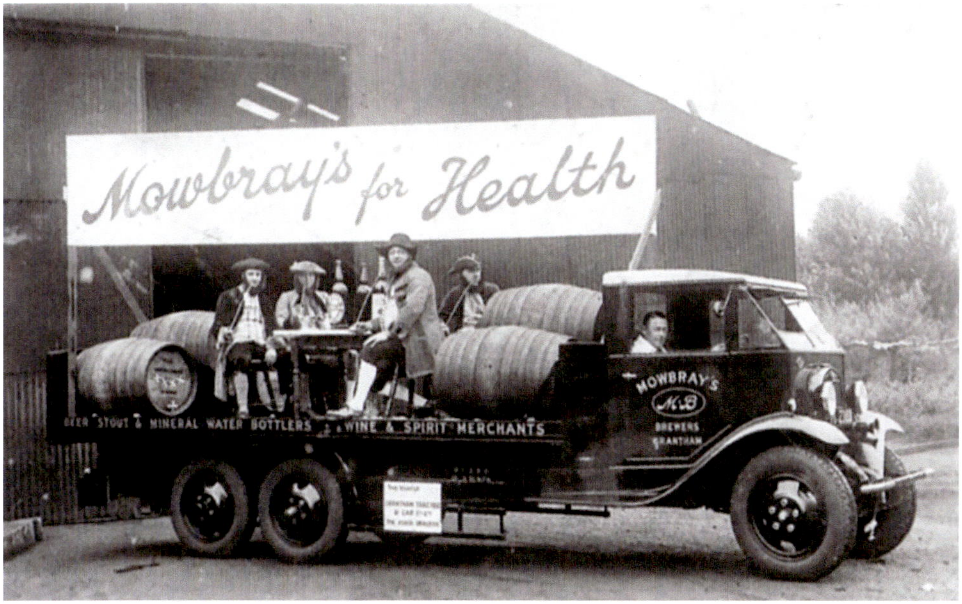

Displaying perhaps a controversial claim for the benefits of Mowbray's beer, this brewery dray is probably about to take part in a local event, complete with a quartet of actors. This photograph probably dates from the early 1930s. Note the 'M.B.' logo, which was used by Mowbray's for many years. (Postcard)

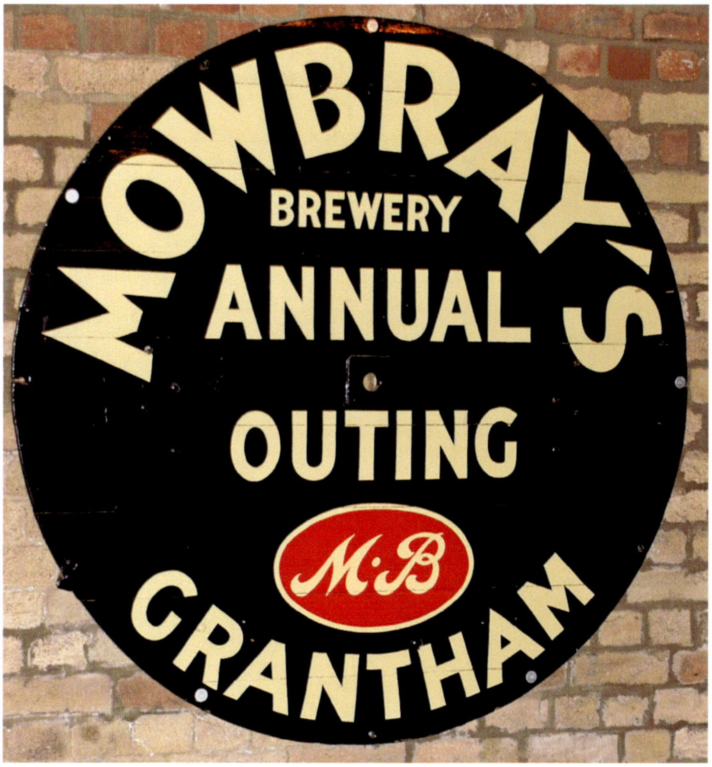

Annual outings, usually to seaside destinations, were a feature of the larger brewing firms, and Mowbray's were no exception, with Blackpool, Scarborough, Southport, Great Yarmouth and Hunstanton amongst those visited in the 1930s. This headboard graced the smokebox of the steam locomotive hauling the train, complete with 'MB' branding in house style. (John Law)

secretary for two years, on the understanding that she would have to resign when a permanent (i.e. male) replacement was found. Despite the patronising treatment, she stayed on in executive roles in the brewery after her marriage to George Schwind in 1934 and became a magistrate.

Both world wars hit the brewing industry hard, with output restrictions and severe labour problems as many staff away on military service. Bomb damage in the Second World War affected Mowbray's badly; Grantham was home to several heavy engineering firms, and several pubs were hit as well as parts of the brewery. The entire frontage of the Joiners Arms in London Road was blown out in September 1940, leaving Mowbray's no option but to demolish the pub, and further afield the Butchers Arms in Eastgate, Bourne, was also completely destroyed when it received a direct hit from a damaged German plane in May 1941. Two of Mowbray's staff sadly did not return from the war, including the assistant brewery manager who was seen as a rising star. Problems also continued at management level with several key staff leaving or retiring. Edward Dunkerton, a magistrate and retired bank manager, became managing director and chairman in 1941, but the firm so lacked experienced brewing knowledge that they invited Harry Bateman, chairman of rival brewers Batemans of Wainfleet, to join the board. He became a director of Soames of Spalding for very similar reasons.

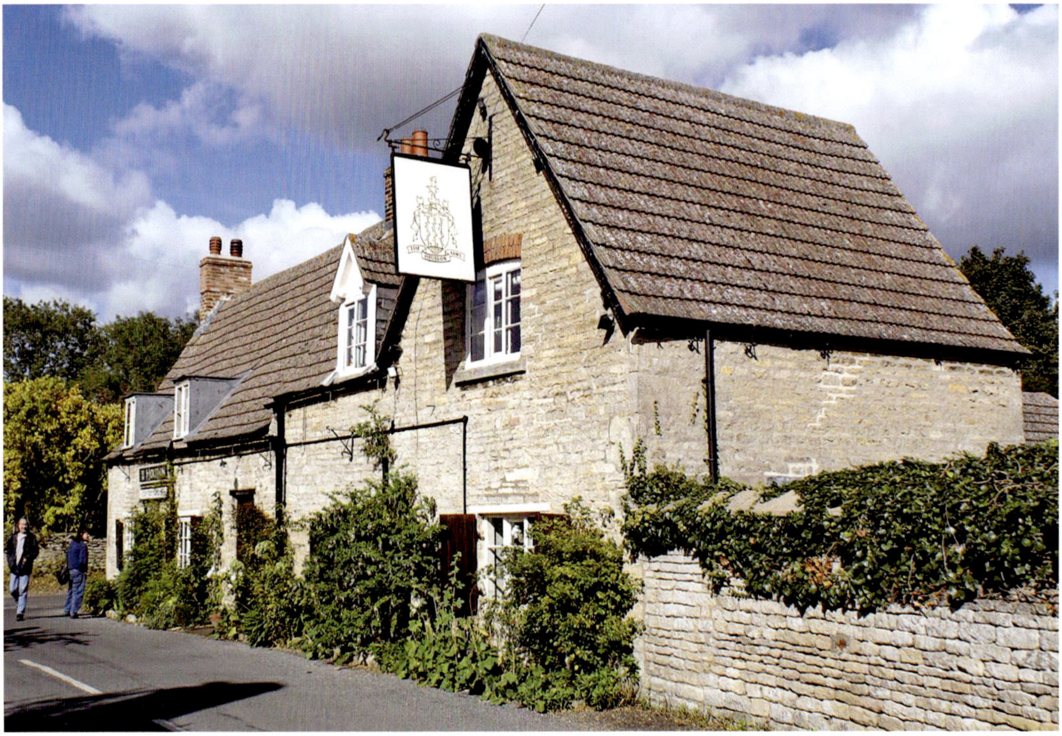

The Houblon Arms, a charming stone-built pub in the village of Oasby, seen here in 2009, was bought by Mowbray's in 1950. The building dates back to around 1650 and the pub name commemorates Sir John Houblon, the first Governor of the Bank of England and a local landowner. (John Law)

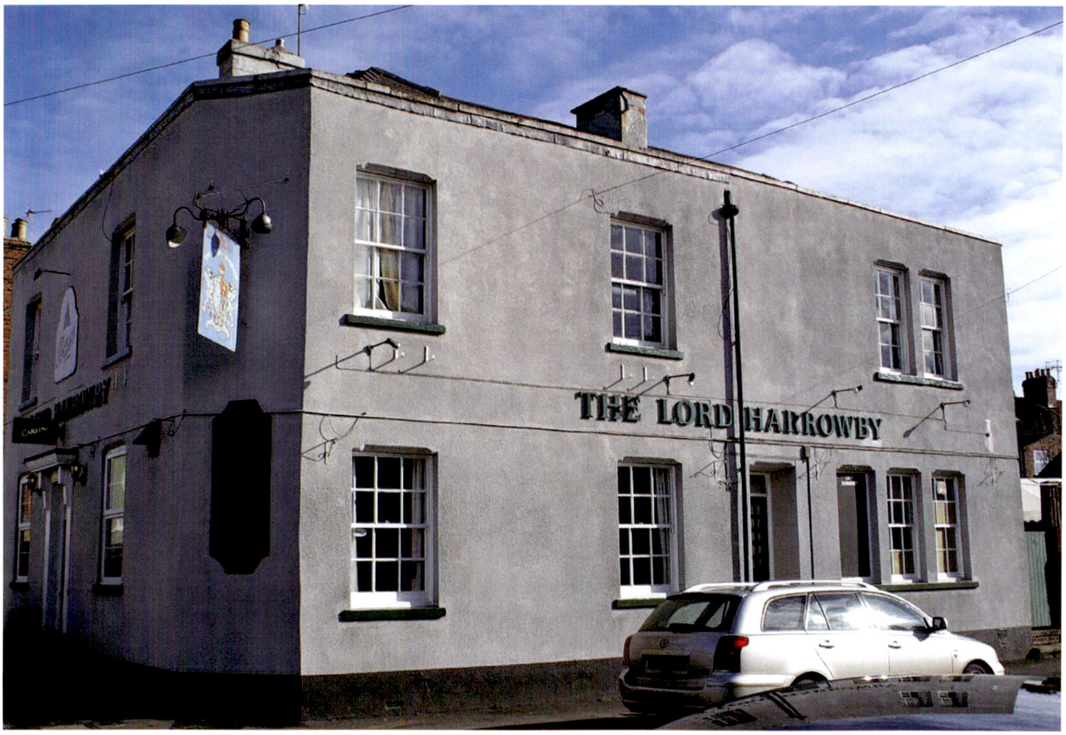

Mowbray's bought the Lord Harrowby in Dudley Road, Grantham, in 1930. Its low-hipped roof is very clear in this view dating from 2009. (John Law)

Mowbray's actually rode out the immediate post-war period quite well, despite the troubles of other brewers. Profits were buoyant, partly helped by Mowbray's having several fairly modern pubs in urban locations. Major investment in the brewery and some of the pubs was needed, however, funded by more share issues which were quickly snapped up; healthy dividends attracted investors, although the existing shareholders managed to retain overall control. As time went on, the outlook began to seem less bright, but it was still a shock when the news broke in February 1952 that Mowbray's were selling out. Dunkerton and another director, Vernon Sedgwick (a Hutchinson descendant) met Bernard Dixon, the chairman of J. W. Green's brewery of Luton, in December 1951 for informal talks that led to a takeover offer two months later in February 1952. Harry Bateman resigned immediately by telegram, citing 'personal matters between the chairman and himself'. Almost certainly there had been a frank exchange of views; Soames had sold out only two years before and now another firm advised by Harry had chosen to lose its independence.

Dixon, a far-sighted and ambitious man who understood the power of marketing, was overseeing the expansion of Green's across much of southern and eastern England. His master plan for the East Midlands was to merge Mowbray's with Soulby's of Alford, already snapped up the previous year, and concentrate brewing at Grantham. This was quickly implemented, along with selling Green's beers in all of Mowbray's pubs – including Dragon's Blood, an early branding triumph for an unfamiliar premium ale. But more turmoil happened in 1954 when Green's bought Flower's of Stratford, choosing to take on the latter's name, so that Flower's adorned pubs across the old Mowbray estate until 1962, when the firm was bought by national corporate Whitbread.

Brewing at Grantham had ceased in 1958, although the Commercial Road continued as a distribution depot for Whitbread until June 1967. Although many of those buildings remain in use as industrial premises to this day, the buildings fronting London Road were largely replaced by a car showroom in the early 1960s – apart from the former Redhead brewery, which was used as a carpet showroom for many years until demolition in 2004. The entire London Road site is now covered by retail warehouses and car parks, adding to traffic congestion and offering a dispiriting prospect to this part of Grantham.

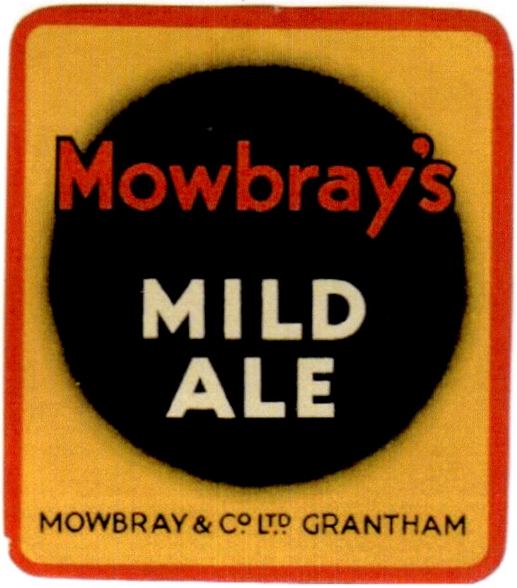

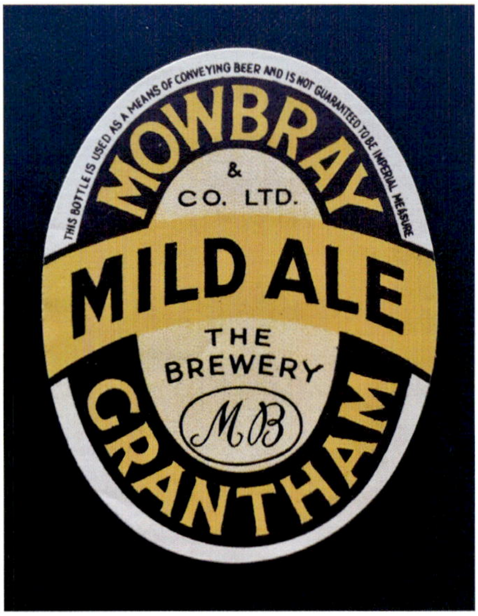

Above left: Mowbray's Mild Ale beermat. (AC)

Above right: Mowbray's Mild Ale label. (AC)

4

Sergeant's of Brigg

Always a small brewery firm, its estate rarely exceeding twenty houses, Sergeant's of Brigg traded for over 130 years in apparent independence until 1967, when its name was summarily dispensed with by brewing giant Bass Charrington. The Sergeant family had in fact sold out to Hewitt's of Grimsby in 1954, but such was the popularity of their beer with Scunthorpe steelworkers that the takeover was kept firmly under wraps.

The firm was founded in the early 1830s by the impressively named Antonie Middlemore Sergeant, the son of a London schoolmaster, who rented a small brewery behind a house on Bridge Street, Brigg. In 1852, he and Edward Sergeant built a new steam-powered brewery further down the street, behind the (now closed) White Hart inn, on the west bank of the Old River Ancholme, using water from nearby springs at Castlethorpe and locally malted barley.

The new brewery consisted of a large three-storey warehouse-style building with a steeply pitched roof containing two more floors, and a large cellar. There was an engine room and chimney at the northern end of the building, which was possibly a later addition. The height of the premises would have dominated the town, although it had virtually no architectural style; perhaps to mask this, the name 'Sergeant & Co' was painted on a white banner across the full width of the southern gable end.

Trading as A. M. and E. Sergeant and Co., they started to build up a small tied estate in and around Brigg, which was then an important market town and administrative centre for North Lincolnshire. The business however remained small scale, competing with the town's other brewers, Sutton, Bean & Co., whose Britannia Brewery in Wrawby Street was built a year after theirs. It remained firmly in family hands for four generations: Antonie Sergeant handed the business over to his sons, John and Antonie. On Antonie Jr's death in 1912, John's son-in-law, James Raynham, took over day-to-day management, with John's son, Scunthorpe solicitor Henry (Harry) Sergeant, having an executive role.

After the First World War, Sergeant's continued their stately progress, achieved through a small but high-quality tied estate, and a proficient head brewer in Tom

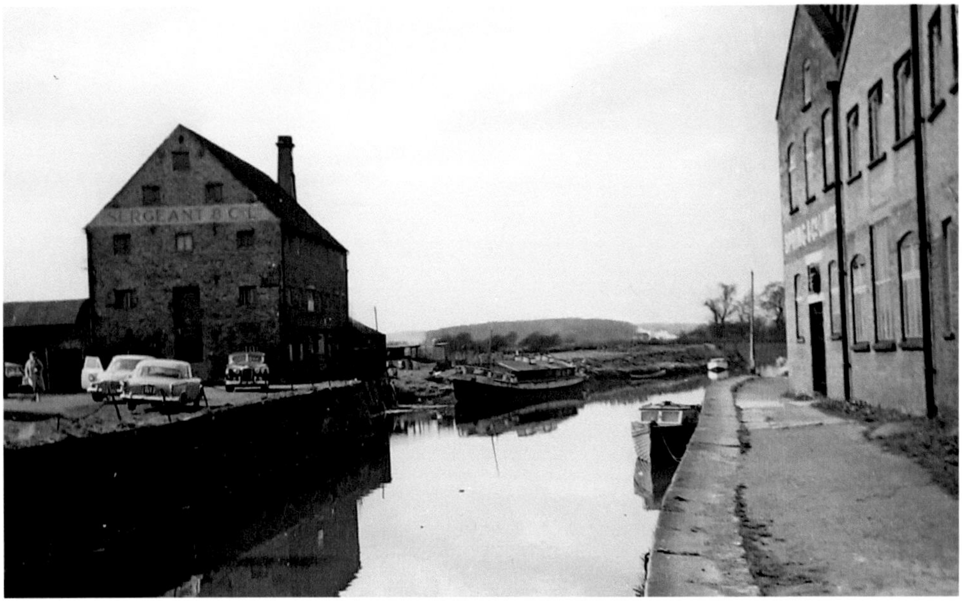

The brewery in 1959 whilst still in use and displaying the Sergeant name, in a photograph taken from the east bank of the River Ancholme looking north. Spring's jam factory is on the right. (John Thompson)

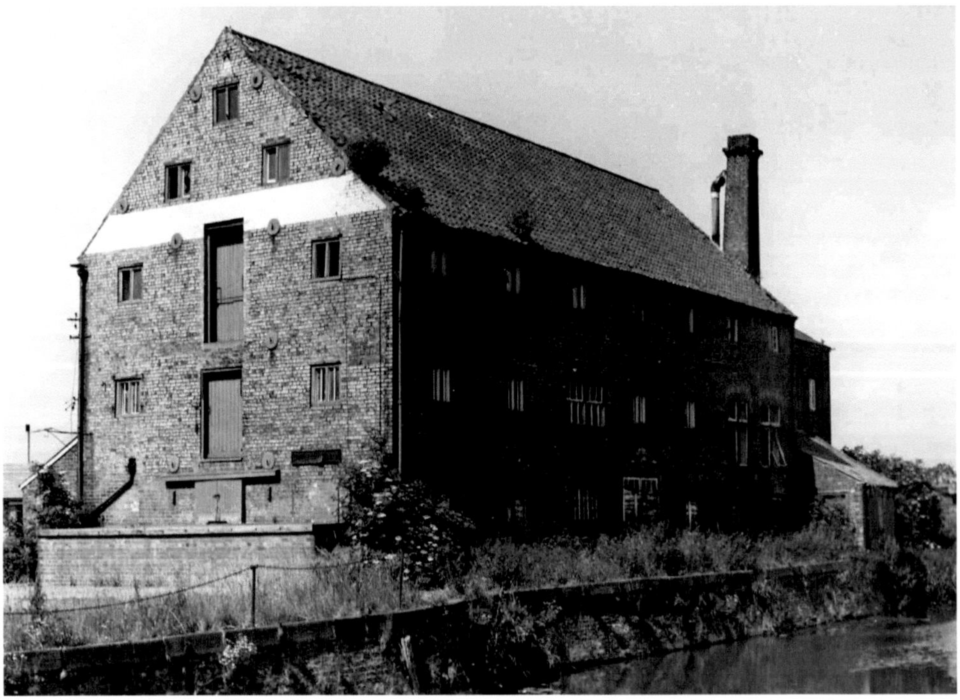

A closer view of the derelict brewery in 1980, with the boiler chimney at the north end and access doors to the first and second floors at the south side. (Brewery History Society)

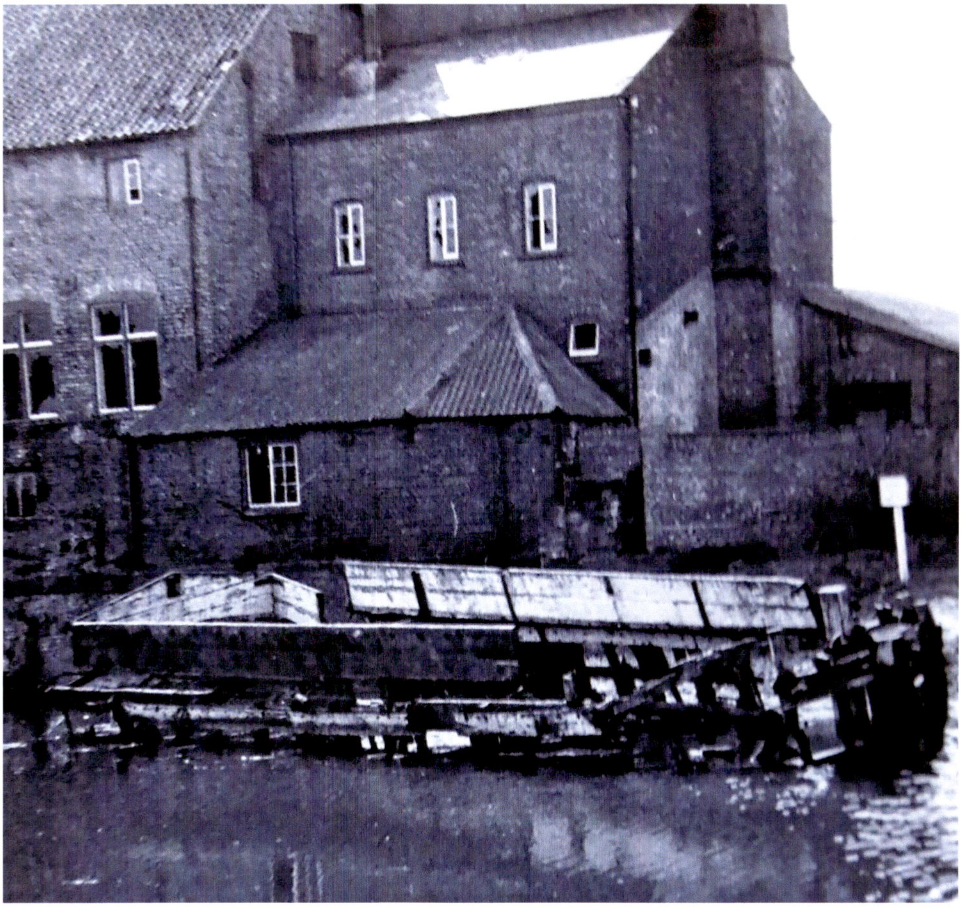

The northern end of the brewery showing the brewhouse, and a beached vessel in advanced disrepair, some time in the 1970s. (Nigel Fisher)

Cheatle, who served the company for around thirty years. The business was converted into a limited company in 1928, with Harry Sergeant and James Raynham providing the modest share capital of £4,000. In the previous year, Sergeant had taken the unusual step of turning his large Scunthorpe home, Wortley House, into a hotel; it was Sergeant's first outlet in the fast-growing town. Their rivals Sutton, Bean & Co. had beaten them to it by over twenty years, their impressive Britannia Hotel having opened in 1906 (Sutton's, who ceased trading in 1924, had a more widely spread tied estate). But steelworkers in the town's workingmen's clubs were quickly acquiring a taste for Sergeant's ales. Dolphin bottled beer was a particular favourite, helping the company gain a first prize at the 1932 Brewers' Exhibition in London.

The sudden suicide of James Raynham in 1931 must have been a great shock. He was replaced by a brewery manager, Walter Lees, with Harry Sergeant handling all major decisions. Sergeant had firm views on public house design, including the provision of a bowling green or tennis court wherever possible, as he had done at the Wortley House Hotel, although the increasing need for car parking spaces soon ended that idea. After several other breweries had opened large estate pubs in Scunthorpe, Sergeant eventually commissioned one of his own, the Dolphin on Messingham Road, which opened in March 1939. It served new housing estates to the south of the town. The bar featured repurposed lounge seating from the Cunard liner *Berengaria*, which had just been taken out of service for scrapping.

In 1954, Hewitt's of Grimsby acquired control of the company and its eighteen pubs, although Harry Sergeant remained managing director. Their reputation for good beer encouraged Hewitt's to keep the two companies largely separate and continue using the Sergeant name. It was Hewitt's investment that allowed Sergeant's to build a further four new pubs: the first was opened in 1957 at Broughton, between Brigg and Scunthorpe, named the Dog and Rat. Although that sounds like it belongs to a 1990s brewpub, the name was inherited from the ramshackle old, whitewashed inn, which the new pub replaced. Three years later, the Ancholme on Grammar School Road became Brigg's first new pub in many years. The Pied Piper in Willoughby Road, Scunthorpe, followed in 1962; breaking the company's traditional designs up to that point, it had a more modern, angular design. Finally, the Wishing Well at Worlaby, a few miles north of Brigg, opened its doors in December 1964 as the village's first pub. Unfortunately, time has not been kind to any of these four, with the Ancholme and

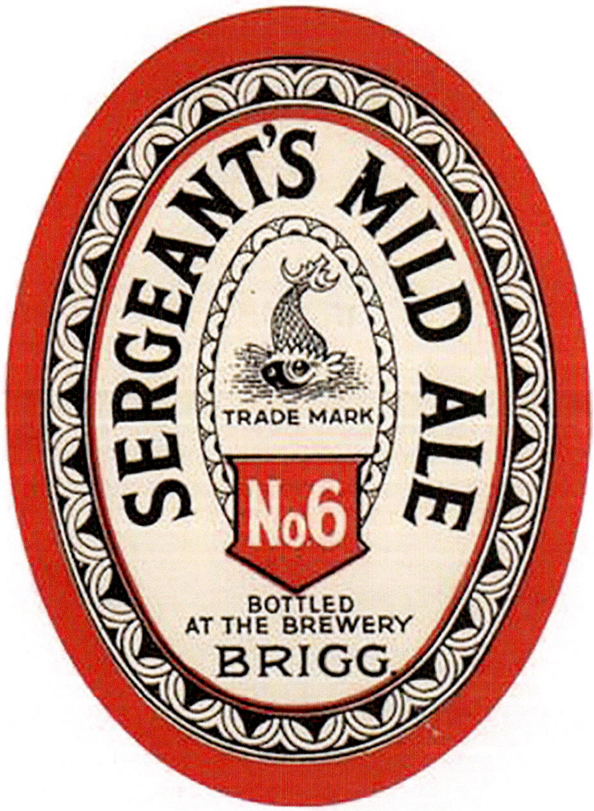

Sergeant's Mild Ale label. (AC)

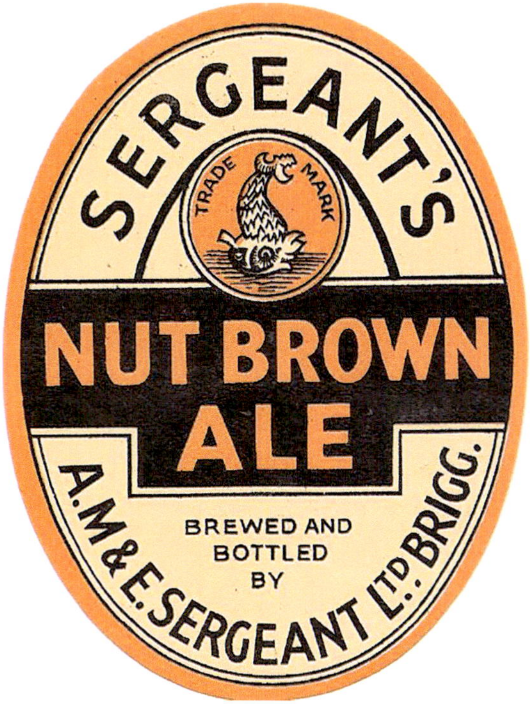

Sergeant's Nut Brown Ale label. The dolphin logo looks pretty drunk in this and other examples! (AC)

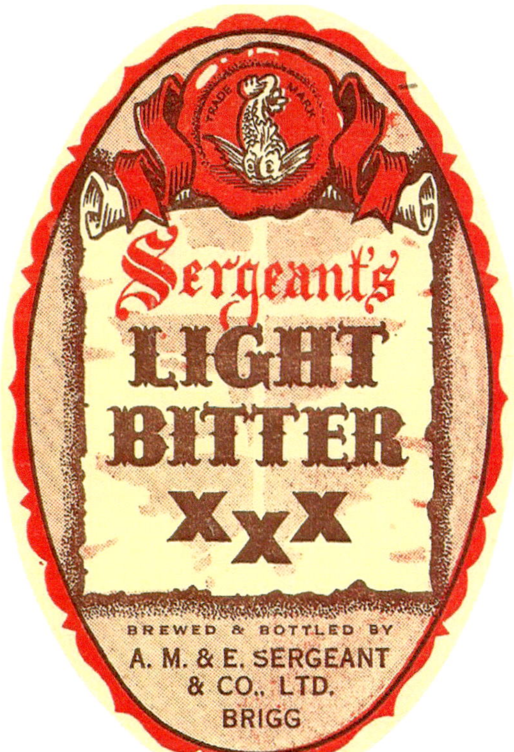

Sergeant's Light Bitter XXX label. The 'Olde English' appearance and use of calligraphy was popular with several breweries in the mid-twentieth century as it presented an image of long-established dependability. (AC)

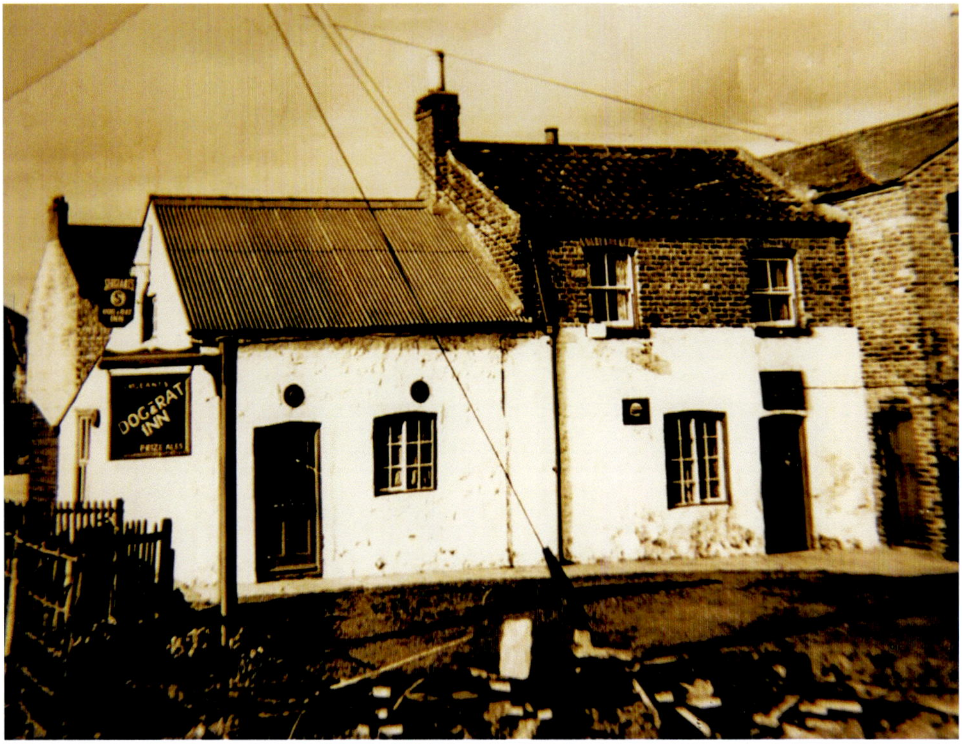

The original Dog and Rat pub in Broughton, possibly Sergeant's smallest pub but boasting two pub signs. Note the tie bars and iron roof on the extension to the left. Why the building was whitewashed to the level of the upper-floor windows isn't clear. (Broughton Heritage)

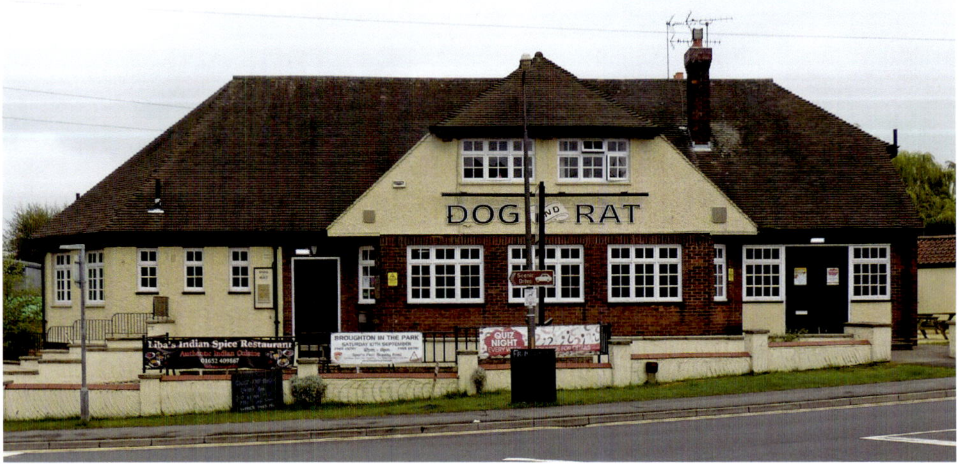

The Dog and Rat in Broughton, between Brigg and Scunthorpe, reopened after its major rebuild in February 1957. As unlike its predecessor as it could possibly be, it closed in 2018 and is seen here a few years previously. (Mat Fascione)

Wishing Well demolished, the Pied Piper adapted rather clumsily as a supermarket and the Dog and Rat likely to be converted into housing.

When parent company Hewitt's became part of the Charrington empire in 1962, the future of the Brigg brewery must have seemed very uncertain, although production at the increasingly antique plant carried on for a few years more. However, an ominous development came in March 1967 when it was announced that output would be cut back at Brigg, and it would only supply clubs from that point. Sergeant's pubgoers would have to make do with Charrington ales from a brewery that had recently been acquired in Tadcaster, 60 miles away in Yorkshire. It was therefore no great surprise when the Brigg brewery closed completely at the end of October 1967. The brewing equipment was removed soon afterwards.

The now empty former brewery premises were sold in 1968 to Spring and Company, a Brigg firm which also traded as Laws Soft Drinks, whose factory was on the opposite bank of the Ancholme. They were interested in using the building for storage, but when a survey was undertaken of the still empty premises in 1976 they were found to be increasingly derelict. It was, however, still possible to discern where the mash tun had been from marks on the floor. The site was later cleared and a modern retail unit, occupied by a carpet retailer, now occupies the approximate area of the brewery.

Sergeant's older pubs have actually fared a lot better, with nine remaining open at the time of writing. A highlight is the Dying Gladiator, on Bigby Street in Brigg, which

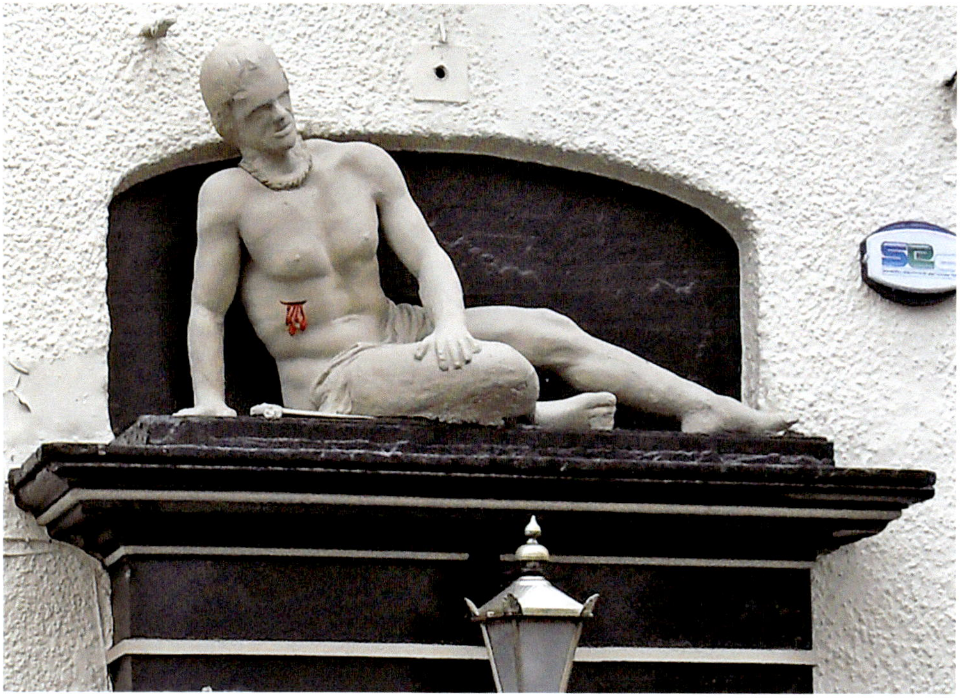

The Dying Gladiator in Bigby Steet, Brigg, was a Sergeant's pub for many years. It still retains this rather graphic feature above the front door, based on an ancient Roman sculpture now in the Capitoline Museum in Rome. (Nigel Fisher)

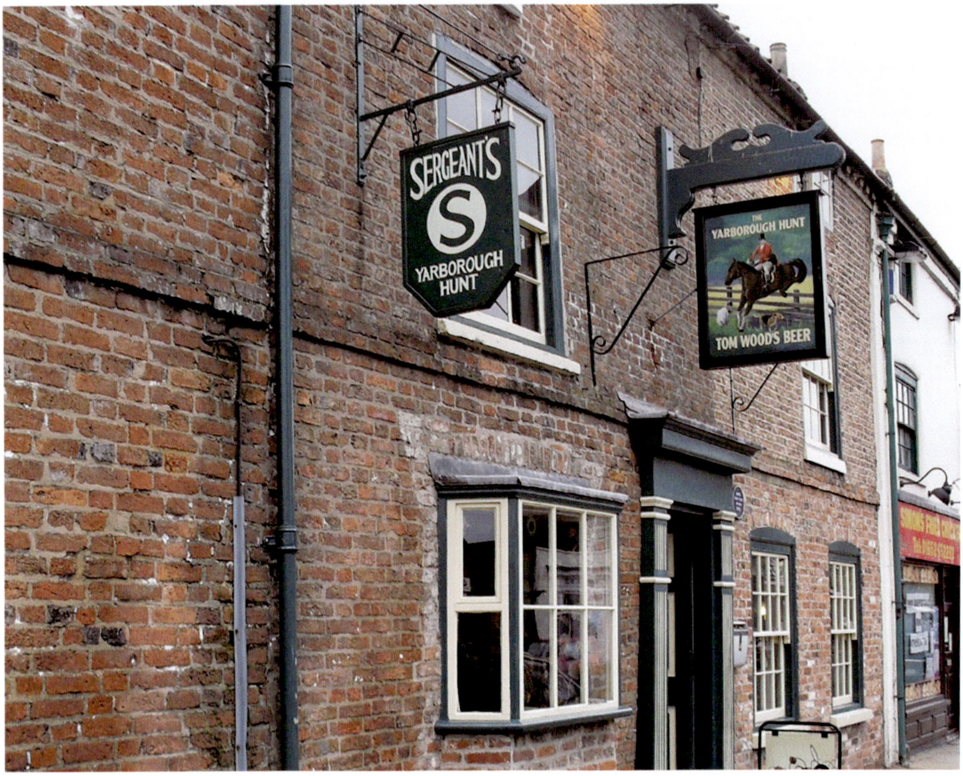

The Yarborough Hunt on Bridge Street, Brigg, was repainted in heritage Sergeant's livery a few years ago, complete with a replica brewery sign. It looks fantastic, and inside retains some Sergeant's memorabilia. (Nigel Fisher)

features above the door just such a figure; it's a copy of a Roman statue, believed to have been made about 150 years ago by a local man called William Clark, whose relatives ran the pub for many years. Harry Sargeant insisted that the sculpture should feature a realistically bloodied wound, so from time to time a dripping red scar was repainted on its abdomen. The Yarborough Hunt, in Brigg's Bridge Street, is also well worth a mention, a pub painstakingly restored some years ago to how it might have looked in its heyday. For a while it even had a green Sergeant's Brewery sign. It's now owned by Lincolnshire Craft Beers of nearby Melton Ross, and is currently their only pub.

Sergeant's beers are still warmly remembered by more senior drinkers, which encouraged a brief renaissance for a few months in 2010. Phil Ellis, head brewer of the sadly now defunct Fulstow Brewery in Louth, recreated Sergeant's bitter, to widespread acclaim. Fortunately, Sergeant's last assistant brewer had saved some old brewing logs which had some of the information that was needed. Described as a 'light, refreshing session beer', amber in colour and slightly sweet, it seemed to go down well in local pubs even though the original Castlethorpe spring water could no longer be used. Sadly, it hasn't reappeared since – but never say never.

5

Soames of Spalding

Once the second biggest brewer in the county, Soames brewed their last pint as an independent company just over seventy-five years ago and are now largely forgotten. That's a shame because in their day they were a large and quite innovative firm with pubs in Peterborough and northern Northamptonshire as well as Lincolnshire, and a history going back to the early nineteenty century.

The roots of the business go back to 1803, when Henry Bugg of Spalding and Thomas Brewerton of Market Deeping, about 10 miles to the south, went into partnership as brewers. Less than a year later, Brewerton was replaced as Bugg's partner by Thomas Sharpe. In 1809 Bugg, by then a sole trader, opened a new brewery in Spalding, choosing a site alongside Westbourne Lodge, a grand and elegant Georgian house on Cowbit Road, looking out to the River Welland. Over the next eighty years the Bugg family became the main brewers in the town, supplying an increasing collection of pubs. In 1846, they snapped up the town's Albion Brewery, in Westlode Street, and most of its estate of twenty-five pubs. They also became prosperous bankers, an elevation in social standing which may have been the reason for a tweak in the family surname in 1877, a change of a single letter transforming them into Burgs. It seemed as if the brewery would pass to a third-generation Burg (or Bugg), but the death of Henry Burg at the age of twenty-seven in 1887 caused a major crisis. Joseph Burg, the young man's father, was left a broken man and determined to sell the business, although he remained living at Westbourne Lodge for the rest of his life, dying there at the age of eighty-eight in 1927.

This was the point when the Soames family rode into the breach. Stephen Soames was a London barrister and JP who had taken up a lease on Cranford Hall, a stately home not all that far away in Kettering. He set two of his sons up as brewery owners: Gerald Soames became the owner of the Anchor Brewery in Long Buckby, Northamptonshire, in 1895, and Francis acquired the Burg brewery in Spalding around 1889. It seems unlikely that either brother had much if any training in brewing, but certainly Francis had a flair for business and leadership. He drove a rapid expansion at Spalding. The Victoria Brewery, a smaller operation a few hundred yards further north along Cowbit Road, was bought in 1892. In the following year they took over the Sleaford

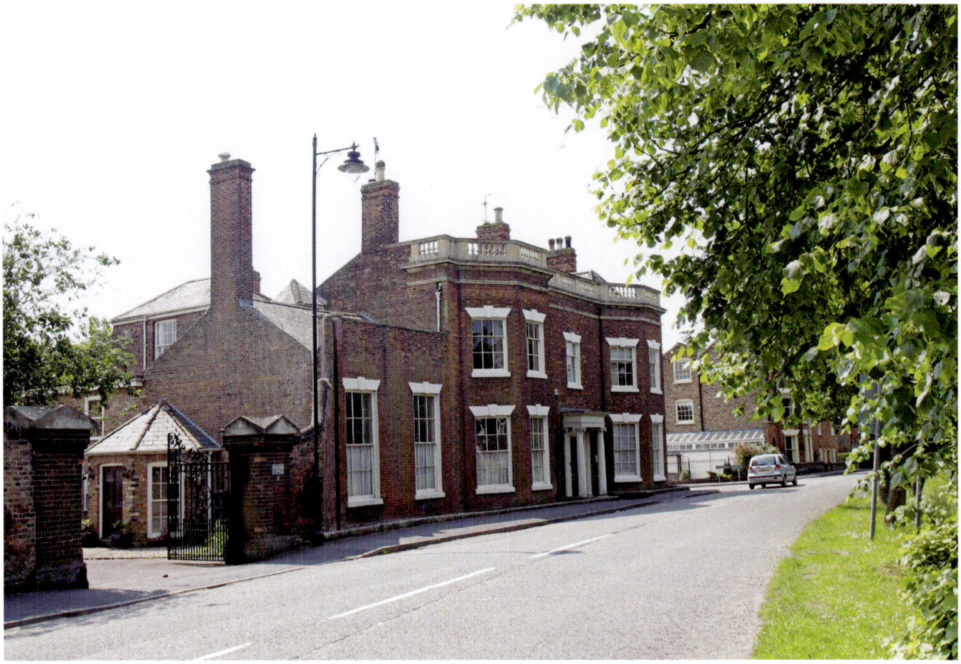

Westbourne Lodge on Cowbit Road, Spalding, looking south. The brewery occupied the space to the right which is now a small housing estate. (AC)

and Market Deeping breweries of John Taylor Marston, who also owned the Dolphin Brewery some 200 miles away in Poole, Dorset (no direct relation of the Marstons of Burton-on-Trent, whose then rapidly growing business is still with us today as a pub company). Soames had financial help to buy the Marston operation from the Peacocks of Greatford Hall, a wealthy Lincolnshire banking family. Marston's pubs were retained, but the breweries were closed and sold. Further pubs in Holbeach, Long Sutton and Boston were soon added.

But tragedy would strike the growing and ambitious business. Francis Soames, an extremely keen sportsman, was riding in a steeplechase at Cartmel racecourse, Cumbria, in June 1903 when he was thrown from his horse, sustaining a fractured skull and lapsing into unconsciousness from which he never recovered. His death at the age of thirty-seven left the business badly affected. Recovery measures included the formation of a limited company in 1909, with retired army captain Robert Soames joining Hugh Peacock as directors, and Bert Walker as company secretary. All the shares were held by members of the two families.

Leadership responsibilities were shared between Hugh Peacock, aged forty-three, and Bert Walker, twenty-nine, with Walker overseeing brewing operations from his home in Market Deeping and Peacock looking after the tied estate and finances; Robert Soames seems to have been largely a figurehead. They had their first major challenge in 1913, when the firm's Boston branch manager, Wiliam Smee, was discovered to have embezzled just over £2,000 from the business – a huge sum then, probably more than ten times his annual salary, and equivalent to about £170,000 in modern prices.

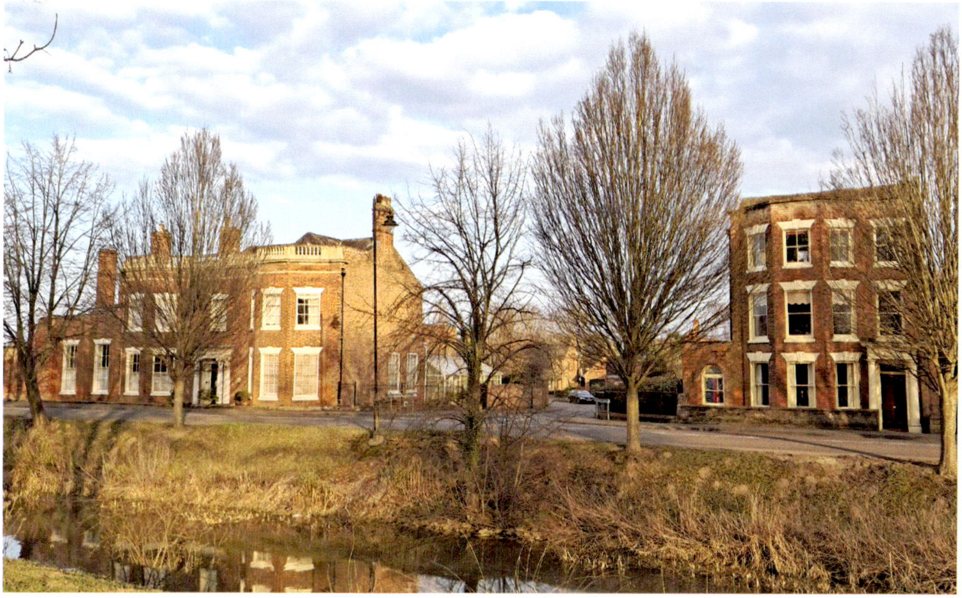

Westbourne Lodge seen from the opposite bank of the Welland on an early spring day in 2025. The gap to the left of the house marks the site of the old brewery, now occupied by housing. (AC)

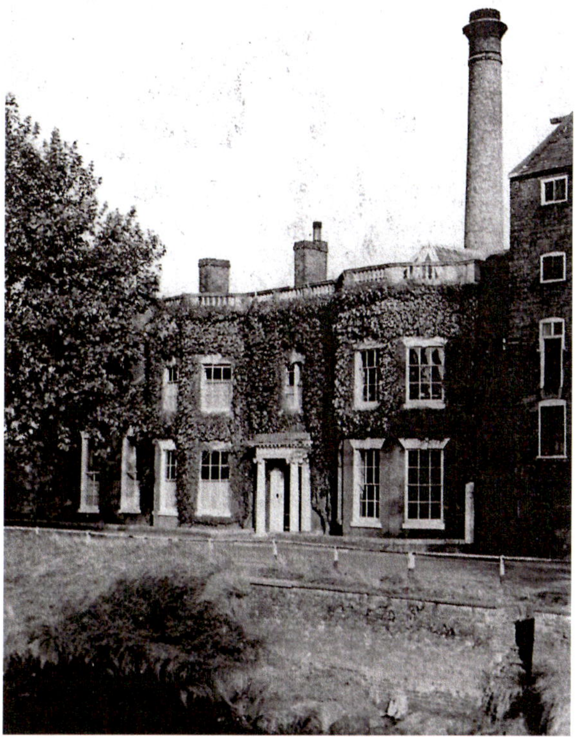

Photographs of the brewery buildings are rare, despite its location on one of Spalding's main roads. This photograph, taken in the late 1940s, was probably intended to depict Westbourne Lodge, but part of the brewery and its chimney can be seen on the right. (AC collection)

Aged forty-seven, he had been with the firm for over fourteen years. The father of five children and a solid member of local society, he was seen as a reliable and trusted employee. But thefts by administrative staff in breweries were not uncommon at the time. Owners such as the Peacock and Soames families had become remote from the day-to-day running of the business, leaving such things in the hands of long-serving employees. Soames only recovered about a tenth of the funds stolen by Smee.

Already Soames seemed like a firm with leadership problems, but those worsened in 1916 when Hugh Peacock died after a long illness aged only fifty-one. Walker's main interest had always been his army career, and when the First World War broke out, whatever time he had given the business quickly evaporated. So the firm made an approach to a local competitor, Charles MacLeod of the Bourne Brewery. He was an able manager in his mid-thirties who also understood the brewing industry. It was the ideal combination for Soames, and MacLeod was willing to oversee the Soames operation for a fee of only £75 over the first six months. He quickly drew up plans to streamline the brewery, proposing the redundancy of several staff; like all breweries, production was severely restricted during wartime. Soames tried to remain loyal to long-serving staff, though, offering them their old jobs back at the end of the war, and taking on a disabled ex-serviceman in their cooperage department.

Perhaps a less welcome return was Bert Walker, who turned up to his first board meeting in many years in June 1919. MacLeod had proven himself a far more effective operator, yet Walker was a director and company secretary who received £800 a year for doing almost nothing compared to MacLeod's increased salary of £200. As a compromise it was agreed that the two men would share £1,000 between them but fortunately Walker was offered 'something in the City' and departed the scene, leaving MacLeod in full control.

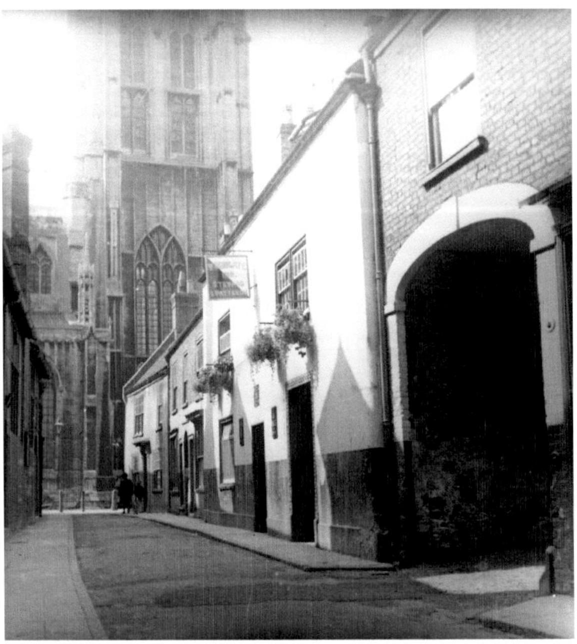

Soames' pub signs were a distinctive white and blue diagonally halved design, as seen on the Wormgate Inn here in Boston, with the Stump, the town's vast parish church, looming behind. This view was taken after Steward & Patteson bought Soames, who chose to continue the Soames house colours. The Wormgate is now known as Goodbarn's Yard. (Roy Hackford)

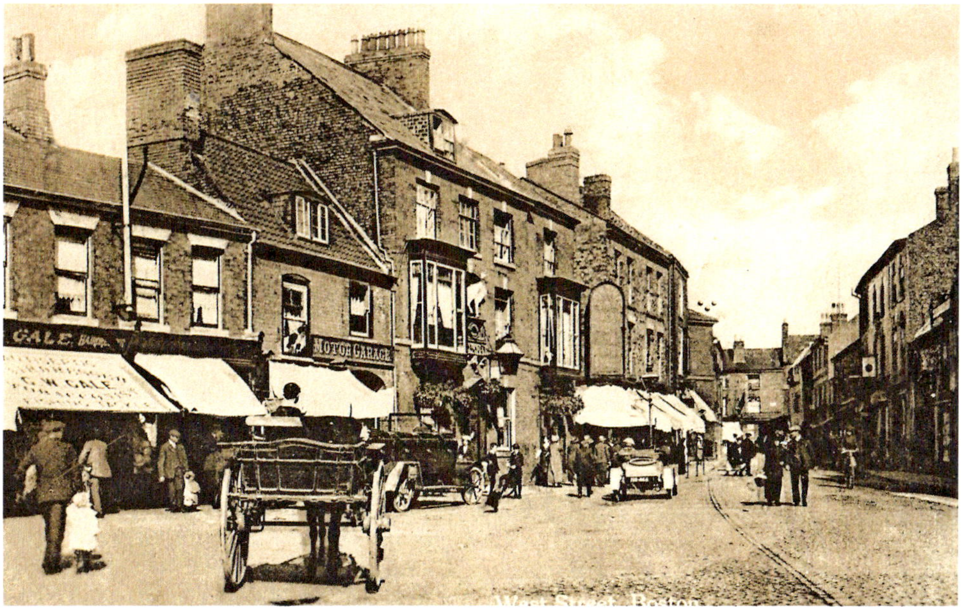

The White Horse was another Soames pub in Boston. The large hostelry in West Street closed in 1957 and was replaced by a supermarket, but a notable survival is the white horse sculpture above the door, just visible on this postcard. It was removed to the gardens of Boston's Fydell House, a grand Queen Anne house which is well worth visiting. (Roy Hackford)

Through steady growth, Soames' pub estate numbered around 170 houses at the end of the war, including Bourne Brewery pubs and other houses owned by the Peacock and Soames families. Several Fenland pubs were bought when they came up for auction, and in 1924 the Rout Green Brewery in Wide Bargate, Boston, was acquired along with the Horry family's wine and spirits trade. Only a few years before, MacLeod had turned down the chance to buy another Boston brewery, Montague Coward's Phoenix Brewery in nearby Grove Street. A visit to the premises had left him unimpressed. The firm remained on the lookout for the right deal. In 1926 the Stamford brewery of George and Henry Hunt, with its seventy houses, came on to the market, but the high asking price of £140,000 and financing problems put them off.

Charles MacLeod remained ambitious, proposing a merger to Mowbray's of Grantham in 1931. His idea made sense; the two firms were of a similar size and although geographically close, their trading areas had only minor overlaps. A combined firm would control nearly 500 pubs. But the talks soon petered out. MacLeod seems to have lobbied for brewing to be centralised at Spalding, a smaller site albeit not as rambling as Mowbray's. And Mowbray's estate was on the whole slightly better situated without the burden of small Fenland pubs serving tiny communities; it's likely they were much less keen on joining forces than MacLeod was. Abortive talks also took place to buy Yates' brewery at Wisbech, and the mineral water business of Lee & Green of Sleaford – a firm whose glass and stoneware bottles remain very popular with collectors to this day. When Lee & Green fell into voluntary liquidation in 1934,

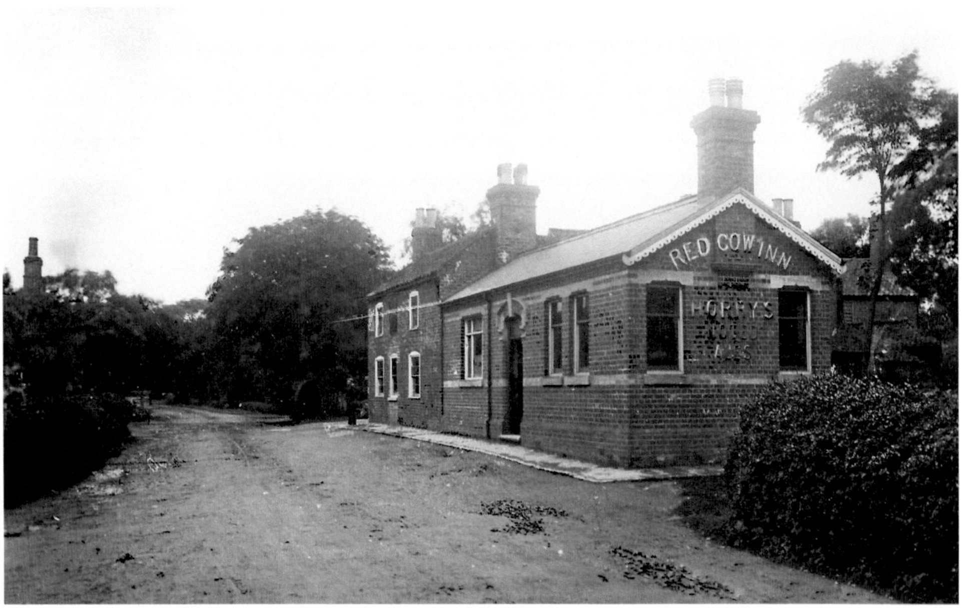

When Horry's brewery in Boston closed in 1924, one of the pubs that passed to Soames was the Red Cow in nearby Fishtoft, seen here perhaps around 1910. It was sold to Batemans of Wainfleet in 1957 and is still trading, although the extension to the right has been demolished. (Roy Hackford)

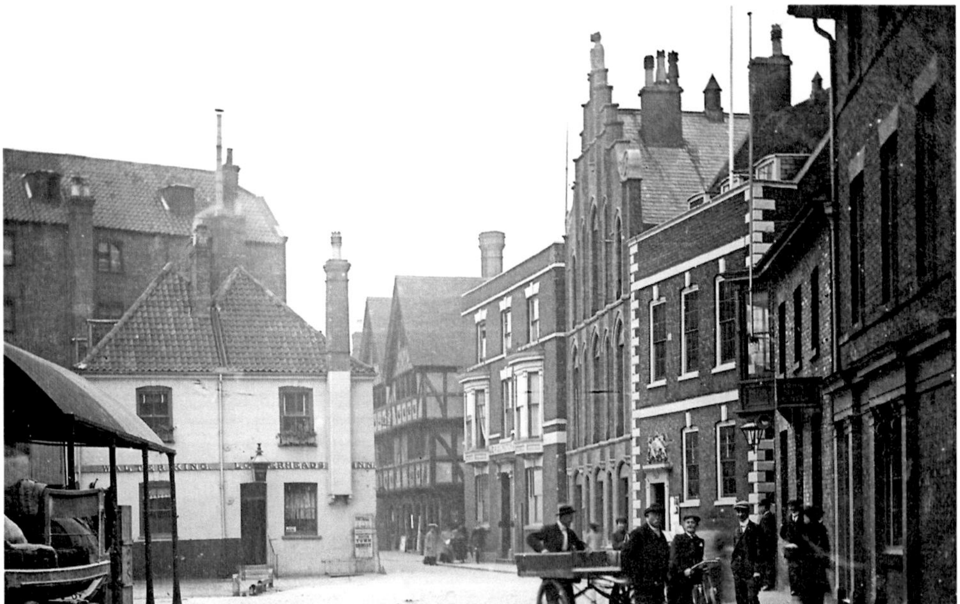

Fairly typical of Soames' town centre venues was the Loggerheads in South End, Boston, another pub bought from Horry's. The major problems it caused to road traffic are very clear in this view from 1905 or 1906. It was finally closed and demolished in 1960. (Roy Hackford)

MacLeod bought the remnants of the business for £15,000, closing and selling off the firm's other factories in Boston, Spalding and Bourne, although their Sleaford factory continued operating until operations were transferred to Soames' brewery in 1946.

The 1930s also saw a modest programme of improvement and rebuilding the tied estate. Soames did not manage to build any entirely new pubs, despite having plans approved in December 1946 for a large estate pub in Atherstone Avenue, Peterborough, which finally opened under new ownership in 1957 as the Halcyon. Similar plans for a pub at Friskney, east of Boston, to attract holiday traffic on the way to Skegness, never saw the light of day. But ten other pubs were completely rebuilt, including the Mill at Boston (now a Batemans pub), the White Horse at Sleaford, and two attractive Dutch-gabled roadhouses at Wrangle (the Coach and Horses, of 1937) and Penny Hill, Holbeach (the Bull's Neck, 1938), both now sadly closed.

The Peacock family had remained significant shareholders in the business, as well as owners of several Soames pubs, and in 1927 a much younger member of the family joined the firm. Hugh Myddelton Peacock, aged twenty-two, had trained as a brewer and initially worked closely with Soames' head brewer, Frank Heard, before joining the Board of Directors in 1930. He enlisted for military service in the Second World War as a Royal Artillery major. Initially serving in Britain, his regiment was sent to India in 1941 and the following year it was caught up in the fall of Singapore, where Peacock was a battery commander. Captured by the Japanese when the garrison fell, he spent over three years as a prisoner of war, made like many of his regiment to work as forced labour in extreme conditions to construct the Burma Railway, the inspiration for the film *Bridge over the River Kwai.*

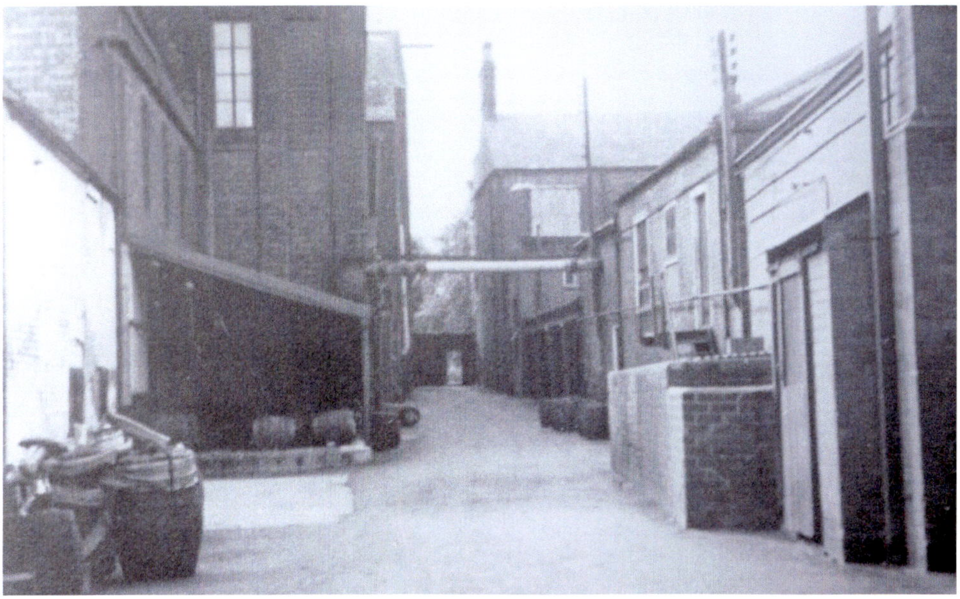

Views of Soames' brewery are very rare, so we have included this slightly grainy photograph taken looking towards Cowbit Road with the main part of the brewery on the left. The rear of Westbourne Lodge is to the right. (Michael Elsden)

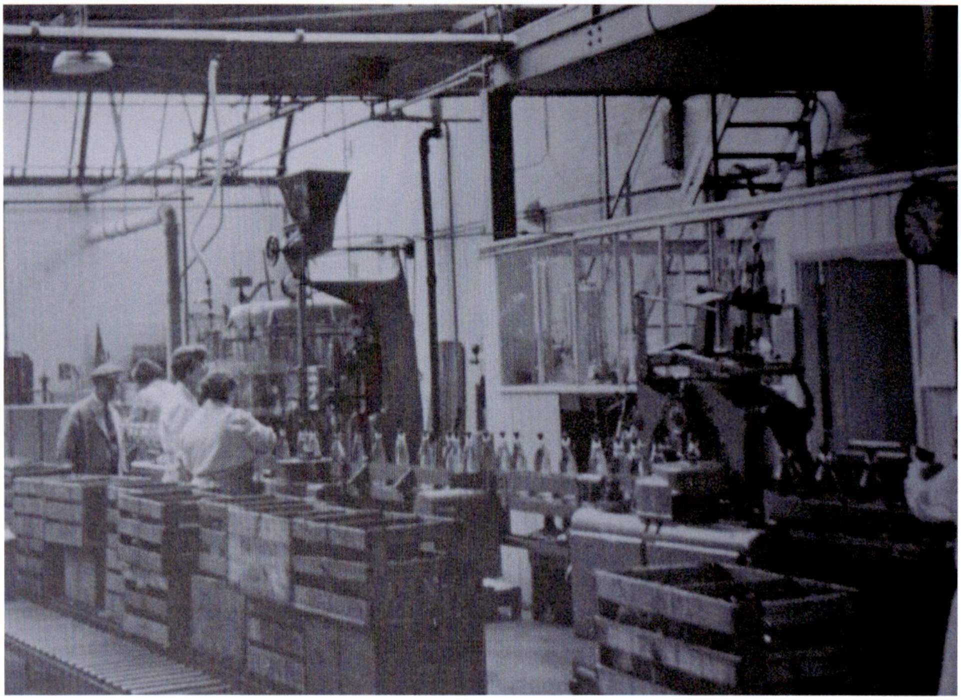

A busy scene inside the bottling room at Soames' brewery, with several women hard at work as a male foreman in a cap looks on. (Michael Elsden)

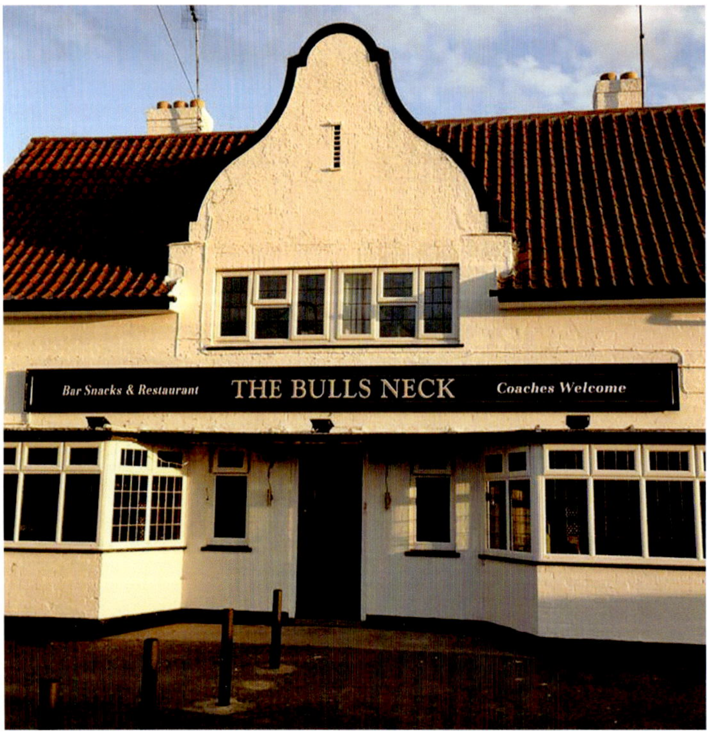

Part of Soames' interwar building programme, the Bull's Neck opened in 1937, cunningly sited on the old A17 at Penny Hill near Holbeach to attract holiday traffic on the slow and winding journey to Norfolk. Road improvements in the 1980s saw the main road diverted to the south although the pub remained open until fairly recently. The Dutch gable design was also used at the Coach and Horses at Wrangle, an earlier loss. (AC)

The firm now faced another leadership crisis. Ward Soames, the final family member on the board, had died in 1941; Peacock was a prisoner; MacLeod, now in his sixties, wanted to retire, and company secretary Leopold Harvey was ten years his senior. Of the remaining staff – those who had not been called up for military service – several were in failing health too. In January 1944 the firm therefore invited Harry Bateman, the energetic chairman of local brewers Batemans of Wainfleet, to join the board.

It seemed that Soames were looking forward to a more stable post-war future, especially when Peacock returned from the war, but in addition to persistent austerity measures, brewers were hit by a storm of decreasing demand, output restrictions, shortages of raw materials and increased beer duty. By the summer of 1946 the firm were having informal discussions with both Mowbray's of Grantham and Soulby's of Alford about 'an understanding' – in other words, some way of sharing production to cut costs. Around the same time Soames were approached by a much larger brewer, Steward & Patteson of Norwich, with an offer to buy them out. None of those talks bore immediate fruit, but a couple more years of hard slog against 'the attitude of the present Government towards brewers' – in the words of Soames' board – made their mind up, and they decided to sell out to their Norfolk suitors. S&P's offer valued the Soames business at a little over half a million pounds, or around £90m in modern terms, for which they would acquire a fairly profitable business, 240 pubs, a wine and spirit trade and a rather ramshackle brewery. The deal gave S&P pubs across South Lincolnshire, Peterborough and northern Northants, all areas where they had no existing presence, but it has to be said that easily half of Soames' pubs were small and increasingly uneconomic, selling less than a barrel of beer a week. S&P already had many pubs like that across rural Norfolk.

For the merger to work in a difficult market, economies were inevitable, the main one being the immediate closure of the Spalding brewery, the disappearance of Soames' beer range and the loss of several jobs, although the site was retained as a distribution depot for many years. The main winners were the Soames shareholders, who received very good value for their holdings.

Under S&P ownership, many low margin Fenland pubs were closed and sold, especially during the 1950s and 1960s, with the proceeds partly reinvested in building a handful of new houses. After the Halcyon pub, planned by Soames, finally opened in Peterborough in 1957, the following year saw another large S&P estate pub, the Woad Man in Boston, open its doors. In 1961 the existing Fishermans Arms, situated in fairly unpromising flatlands at Pode Hole, outside Spalding, was replaced by a huge new pub in the hope of gaining summer fishing trade. It has not survived. All of these pubs were branded 'S&P' from the start, but sported pub signs in the distinctive Soames half white and blue diagonal design – S&P did not impose uniform corporatism.

The former brewery site was closed in 1985 by Watney Mann, who had bought S&P. The remaining buildings were progressively demolished. Only Westbourne Lodge, the grand Georgian building fronting Cowbit Road, which was used as a home by the Bugg family, now remains; its iron gates fronting the road carry S&P monograms. A small housing estate, Westbourne Grove, marks the old brewery site. A number of Soames pubs survive, some of which still feature some reminders of the old company.

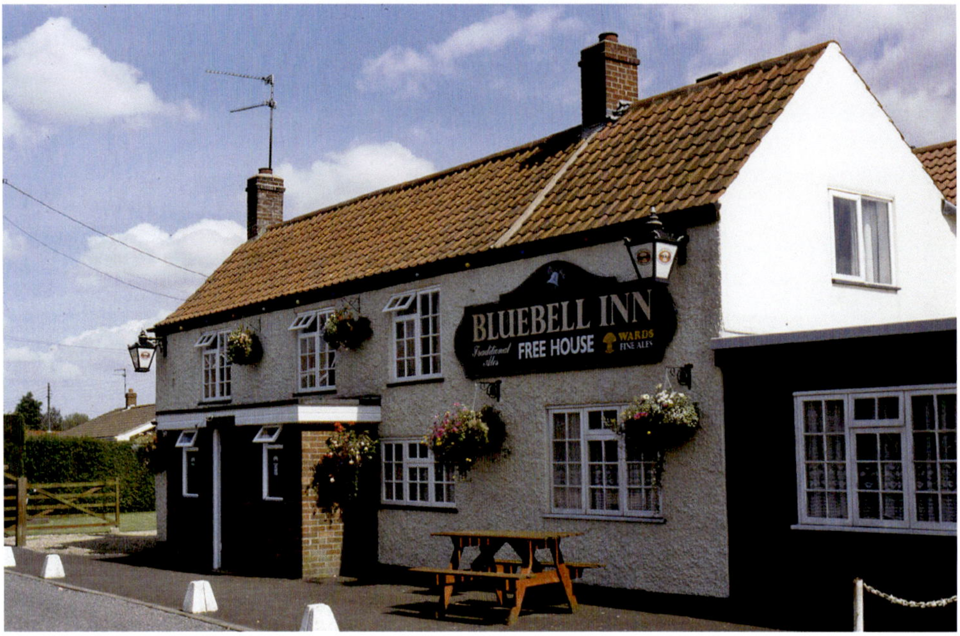

Another of Soames' many Fenland pubs was the Bluebell at Whaplode St Catherine, seen here in 1999 as a free house. It continues to thrive in a relatively remote location, with the added attraction of the on-site Bluebell Brewery, established in 2000. (John Law)

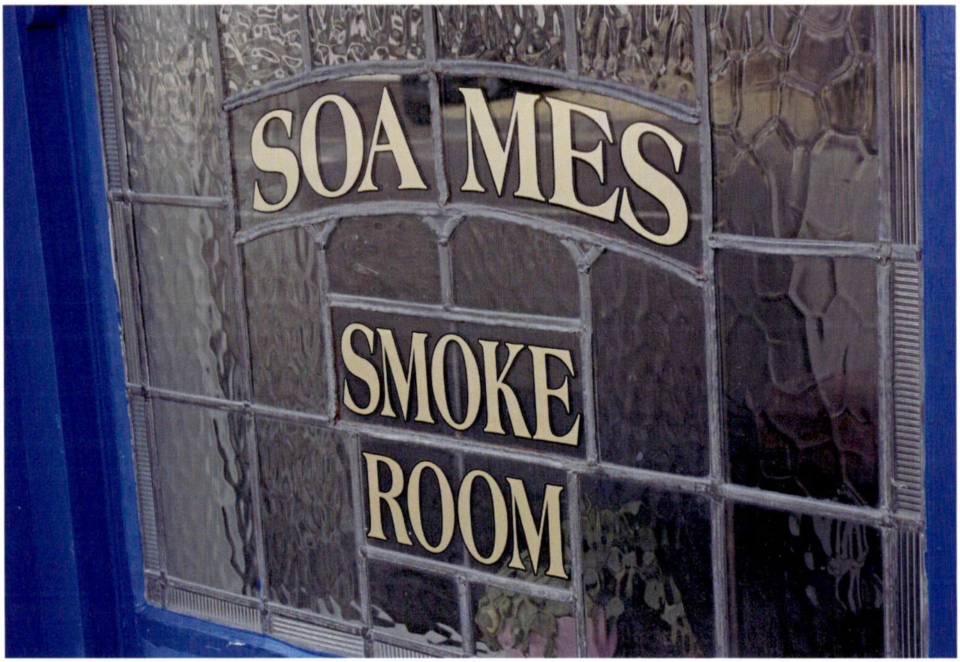

A distinctive Soames relic was the lettering on the pub windows at the Robin Hood in Bourne Road, Spalding, now sadly lost through the conversion of the pub to housing. (AC)

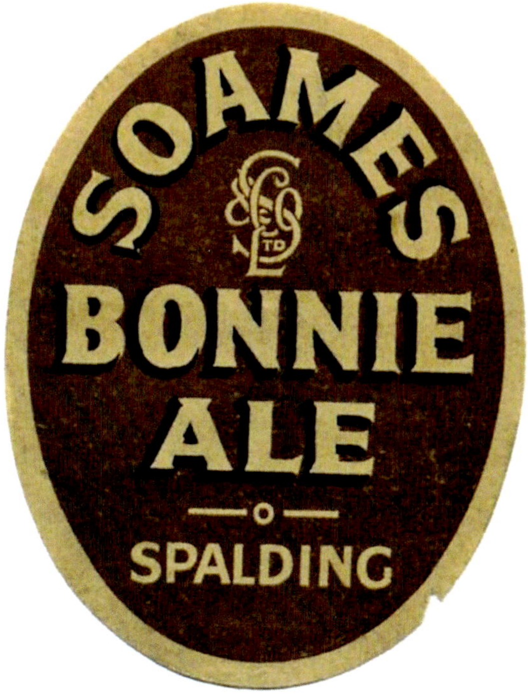

Soames' Bonnie Ale label, a bottled beer whose name probably had something to do with Charles MacLeod's Scottish heritage. (AC)

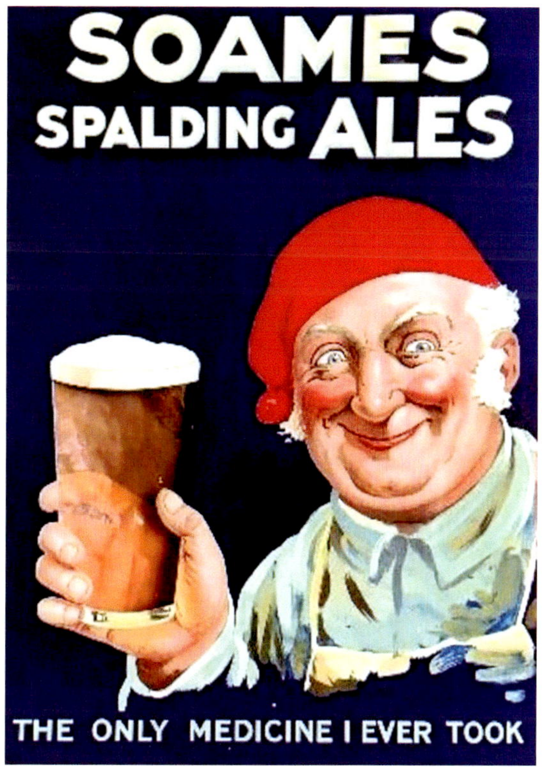

Comedic value in this Soames advert, linking long life and good health to drinking Soames' products. The use of the previous tense was perhaps a mistake though. (AC)

6

Soulby's of Alford

A few miles inland from the North Sea, as the flat country starts to rise towards the Wolds, lies the small market town of Alford. You really wouldn't know to look at it today, but Alford was once home to the bustling Anchor Brewery and maltings, serving a tied estate of nearly 200 pubs and providing employment to many local families. Soulby's – latterly Soulby, Sons and Winch – traded for nearly 120 years, but were swept away in the 1950s, caught up in the brewing industry's belief that big was best, which was to be achieved by hoovering up medium-sized local breweries in the assumption that pubgoers might grumble a bit at first, but would quickly get used to having national beer brands foisted upon them.

Edward Soulby founded the business in 1840, at the age of twenty-six, by taking over a recently built brewhouse in the Fenland township of New Bolingbroke, 7 miles north of Boston. Like many early brewers, he had barley malting, selling coal and brickmaking as other trades to fall back on and so hedge against seasonal downturns. In 1850, he took his brother John Soulby into partnership and acquired the late Thomas Babington's brewhouse in Coningsby, a larger village a few miles further north. The New Bolingbroke brewery continued to operate, however, and it was there in 1854 that the distressing suicide of John Soulby took place, apparently believing that his wife was having an affair with a clergyman from her native Norfolk. Sadly, it was not the only case of extreme depression and anxiety that would affect this family.

Edward Soulby slowly expanded his business in the 1860s, acquiring a few country pubs, and opening an agency in Boston, by some distance the largest town in the area. His biggest decision came in 1869 when he bought a handful of pubs and a small brewery in West Street, Alford. Brewing on the site, tucked away behind the Half Moon inn, dated back to the late eighteenth century. Again, Soulby had spotted an opportunity presented by the death of the previous owner, in this case local farmer Thomas Bradley. Soulby invested £3,000 – then a huge sum – to build a five-storey brewery on the site, with equipment supplied by the London brewery engineers Henry Pontifex & Sons. It was a serious show of intent which would have required major funding, and it's not clear how Soulby found the money beyond the sale the following year of the original New Bolingbroke brewery.

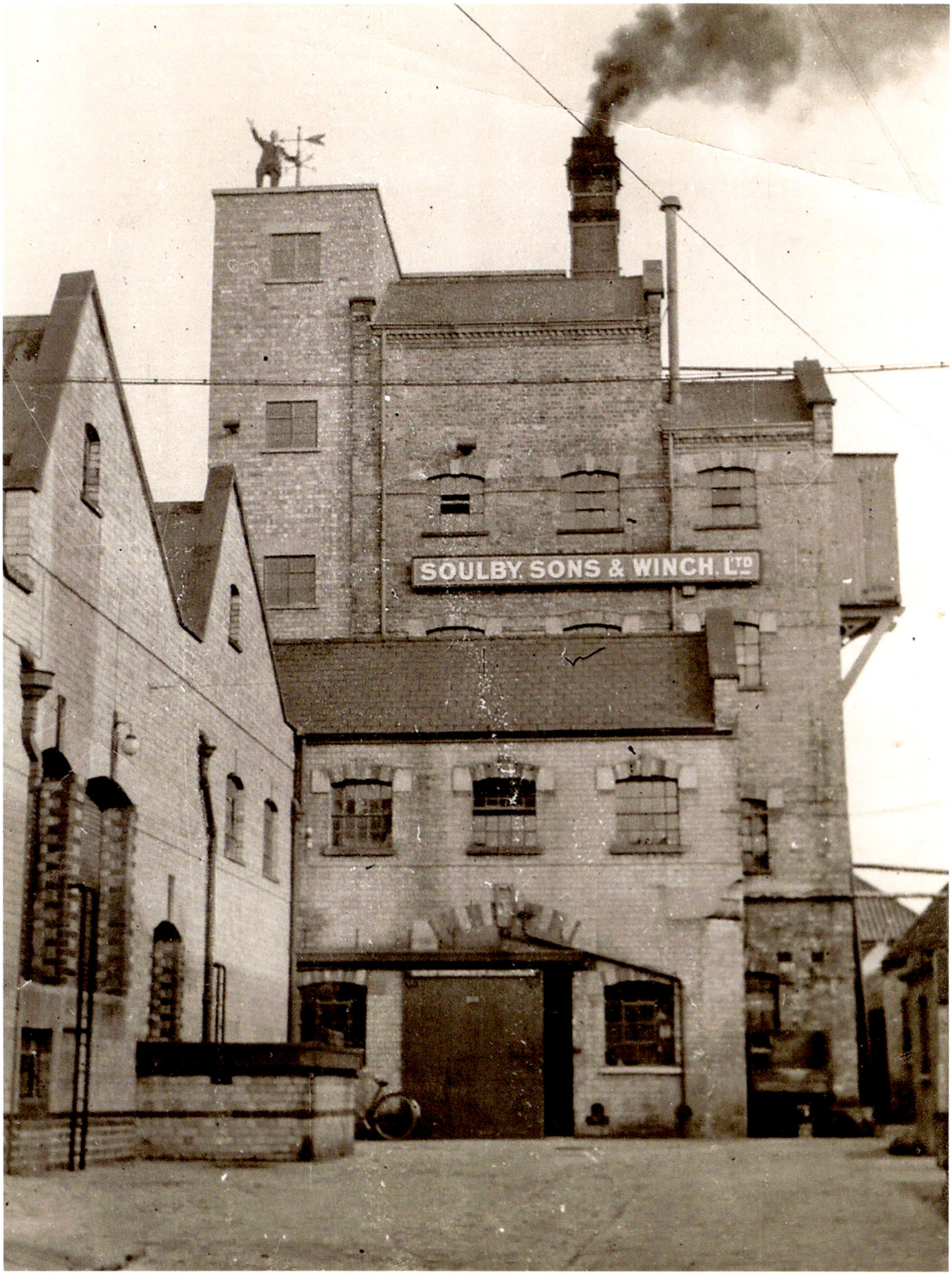

An excellent view of the brewery in the last years of Soulby's, with the chimney dramatically belching smoke. The waving figure on the brewhouse roof and the very tidy yard suggest this view was specially arranged. (AC)

He took his two sons, Edward and Frederick, into partnership in 1870, thereafter seemingly living the life of a successful country gentleman and taking a back seat in the running of the brewery. But there was a reason for that: according to the family solicitor, in his sixties Soulby 'went out of his mind and had to give up the business'. He died aged seventy-six in 1890, but the family's troubles were not over. The passing of Frederick's wife Ada in 1895 triggered a deep depression which led eventually to him drowning himself in the River Nene at Peterborough. That only increased the strain on brother Edward, who was admitted to Ticehurst hospital, Kent, in February 1897 after a suicide attempt involving a 40-foot jump from a Brighton hotel window. Ticehurst was one of the most highly regarded mental health institutions in the country; its patients came from very wealthy families. Edward lived for only a few years more, dying of a heart attack in 1903 aged only fifty-five.

These distressing events unsurprisingly caused a leadership crisis, solved by a merger in 1896 with a young brewer named Tom Winch, the son of a prosperous brewing dynasty from Chatham in Kent. Aged only twenty-three, Winch had bought Thomas East's small brewery and eleven pubs in Queen Street, Louth, earlier that year, no doubt with his family's financial support. He moved quickly to cement the deal, forming Soulby, Sons and Winch Ltd in December 1896. The new company inherited seventy-two pubs, five malthouses and a healthy balance sheet from Soulby's, showing that despite the family problems, business remained buoyant – in fact, Edward Soulby left personal estate valued at over £4 million in today's prices. Under Winch's stewardship, expansion was key: a selection of small breweries in Old Leake (near Boston), Sleaford and Alford were bought and closed in quick succession, with brewing ending at Louth within a few years as well. Concentrating production at Alford made the business more efficient and enabled investment in the tied estate, which grew swiftly beyond 150 houses and pushed into south Lincolnshire. The most spectacular development was at Alford, where a stylish new maltings was built to the rear of the brewery. Opened in 1904, with machinery and plant supplied by Robert Boby Ltd, a prominent engineering firm from Suffolk, the efficiencies that the large, modern building offered allowed the closure of the firm's widely scattered collection of much older malthouses.

Control of the firm seeped away from the Soulby family; by 1907 more than half of the issued shares were in the hands of Tom Winch and his father. In 1899, the extended Winch family had also bought a brewery in Biggleswade, Bedfordshire, which traded as Wells & Winch, and another in Maidstone, creating Style and Winch Ltd and enabling the closure of the old Chatham premises. However, the First World War created problems across the brewing industry, as increasingly men were called up for military service. The widely spread pub estate which Tom Winch had fostered, featuring many houses with limited turnover in remote country locations across Lincolnshire, was difficult to service, so he approached Soames of Spalding in 1917 in the hope of sharing the supply of beer to their pubs, many of which were in overlapping territory. Soames turned him down, but Winch remained persistent and in 1924 he suggested to the much larger regional brewer Marston's of Burton-on-Trent that they should buy him out, as well as Mowbray's of Grantham.

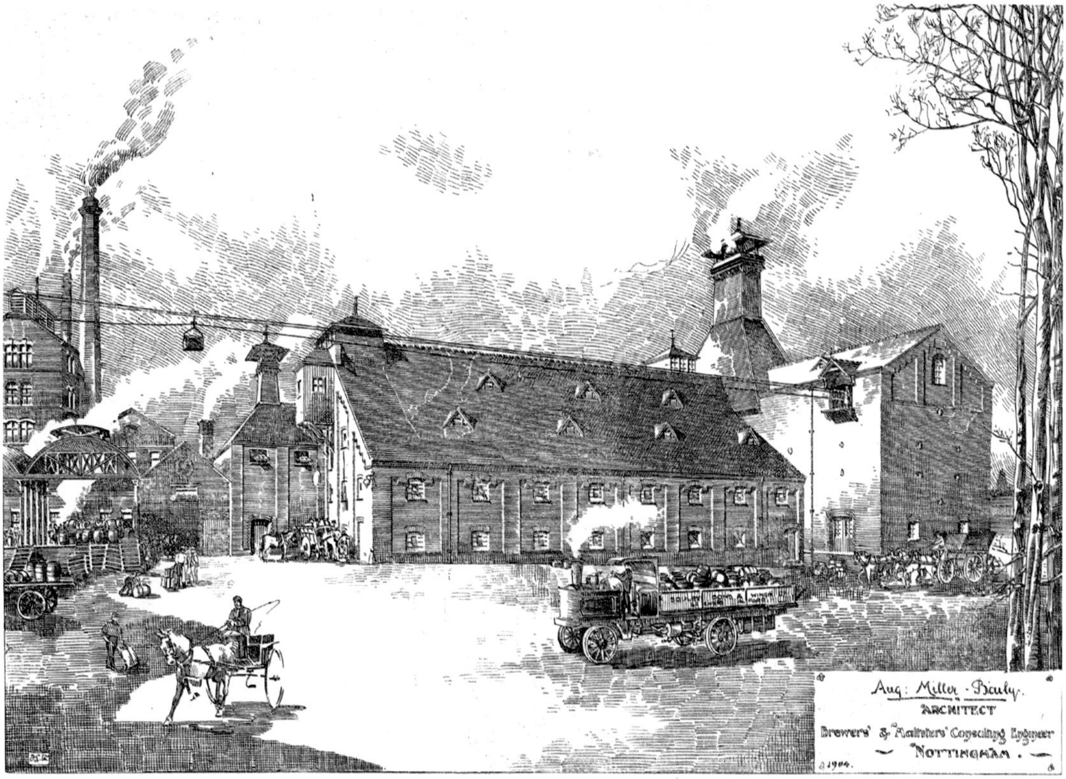

NEW MALTINGS FOR MESSRS. SOULBY, SONS & WINCH LTD., THE BREWERY, ALFORD, LINCS.

Brewery companies and their architects loved to publicise new additions to the brewery estate, often exaggerating features and size for dramatic impact. This drawing by Augustus Miller-Baily, architect of the Alford maltings, dated 1904, is quite close to reality, including Soulby's steam lorry and an on-site winch system. Miller-Baily, who was related to prolific Nottingham architect Basil Baily, seems not to have designed any other buildings, brewery or otherwise. (AC)

Whilst Marston's were not enthusiastic – they'd already decided Mowbray's were of no interest – they did at least carry out some due diligence which paints a useful picture from a hundred years ago. They visited around half of Soulby's pubs, finding them largely well-kept and maintained, but often doing limited business. They were surprised to find pubs such as the Royal Oak in the large village of Ruskington, near Sleaford, sold only two barrels and fifty-one dozen bottles during all of 1924 (it survived until 1963 nonetheless). Soulby's biggest seller, commanding 65 per cent of all sales, was XXX mild, followed by XK light bitter (28 per cent) and finally XPA best bitter (7 per cent). They also brewed a strong stout, sold only in bottles. Although this range largely reflected drinking habits of the time, with lower strength, cheaper beers being the most popular, Marston's were unimpressed by the low profit margins on offer and felt that supplying the estate from Burton, some 100 miles away – for they would have closed the Alford brewery – would be a non-starter. They therefore turned Tom Winch down.

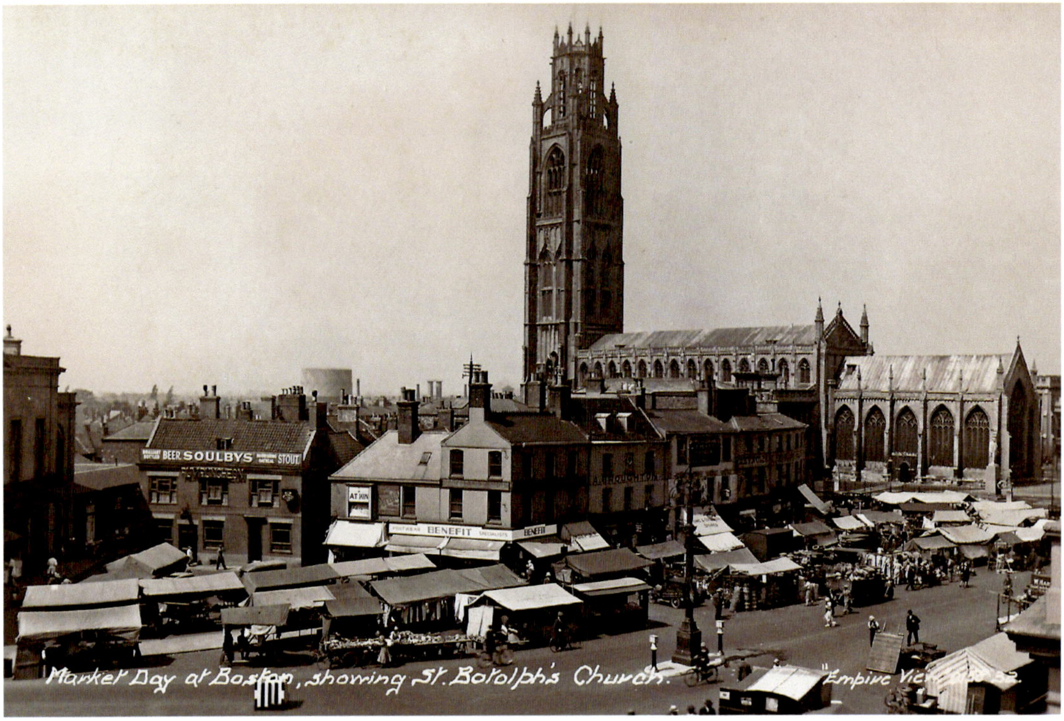

The roof of the Rum Puncheon in Boston's marketplace offered an advertising opportunity too great to resist. Displayed in this postcard view from around 1920 is 'Brilliant bottled BEER. SOULBY's Nourishing Oatmeal STOUT'. The message changed over the years, as other postcard views show. Under its modern name of the Stump and Candle, the pub is now the last one open in the marketplace. (Roy Hackford)

Soulby, Sons & Winch led a quiet existence in the interwar years, with Tom Winch's plans for mergers permanently on hold. In sharp contrast to the family connections with Wells and Winch at Biggleswade – who continued to expand, and rebuilt several pubs in the 1930s – Soulby's had a far more modest programme, with their only entirely new pub, the Hessle House in Boston, which opened in 1936 to serve a housing estate, being a conversion of a domestic property. One other major expense was a full rebuild of the Queen at Donington in 1938. There was a chilling reminder of past troubles in 1938 when a boilerman employed by the firm drowned himself in one of the brewing vessels. He was an ex-serviceman whose father had also died by suicide.

Having avoided any serious damage to its pub estate during both world wars, the company's good fortune ran out on the morning of 5 October 1945, when a raging fire took hold in the brewery tower. It took three hours for the fire brigade to quell the flames. Besides serious damage to the roof and upper floors, some of the brewing plant was damaged. For some weeks the company's pubs had to be supplied by other local breweries who stepped into the breach whilst reconstruction work was carried out, paid for by the firm's insurers. The brewery was therefore rather disfigured in later years by an unsympathetic boxy tower at its eastern end.

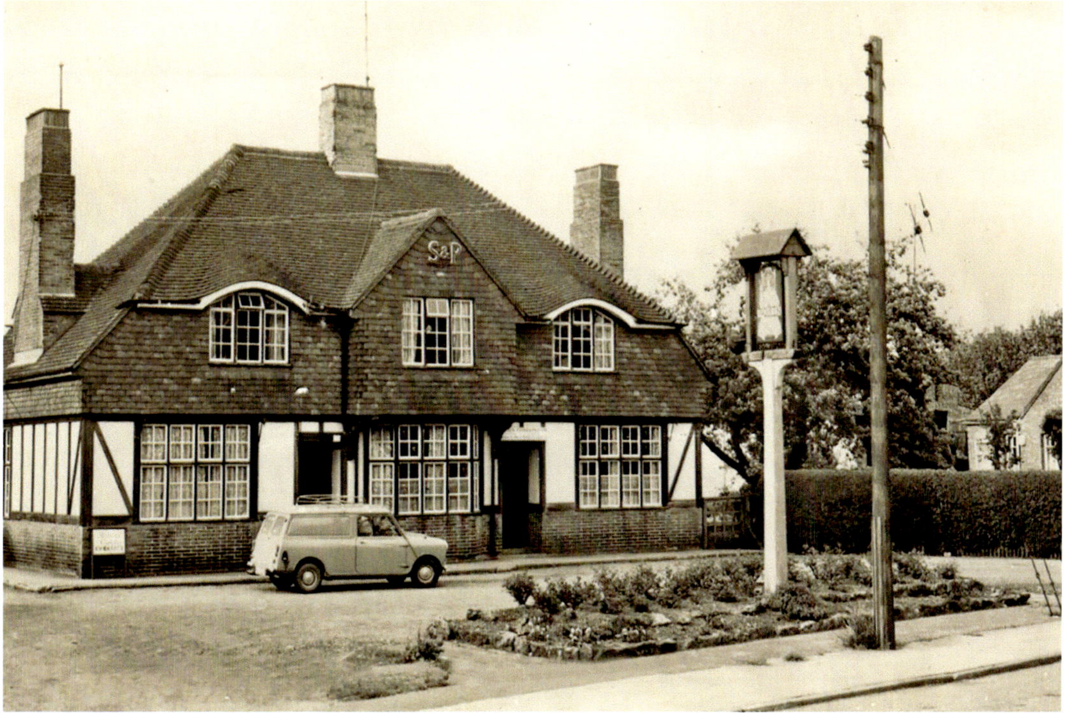

Soulby's rarely carried out any full-scale rebuilding, but an exception was the Queen at Donington, where a rambling old thatched inn was replaced in 1938 by a much larger pub in a 'Home Counties' style with tile hanging and half timbering, and a partly thatched roof. Major repairs and a new tiled roof were needed after a fire in 1950. The pub is seen here in the 1960s having passed to Steward & Patteson of Norwich. It closed in 2014 and is now a private home. (AC collection)

Despite this setback, Soulby's recovered well in the first years of post-war peace, returning healthy profits despite many of its pubs being tucked away in remote Fenlands or small Wolds villages. We know that in the early 1950s, around a third of the firm's 137 pubs were selling less than a barrel of beer a week, with one pub, the Wilson Arms at Walcott, only needing one barrel every month. This dismal picture was offset somewhat by much better performing houses in Louth and Boston, where the Eagle – still popular to this day – averaged 433 barrels a year. Overall, by 1950 profits had begun to decline.

After the death of Tom Winch in 1942, the firm's board comprised a selection of local solicitors and former military men, including his son Cosmo Gordon Montagu Winch – or Colonel C. G. M. Winch, as he was usually known. Former military officers were a popular management choice with corporate boards of the post-war era, who seem to have felt that a chap who served his country could equally well run a brewery, with the use of their former military rank lending, perhaps, the hope of strong leadership.

Mature businesses like Soulby's, with ageing directors, could often be tempted by takeover bids, and that's exactly what happened. Discussions were opened with Green's of Luton, with news breaking in August 1951 that the company was being

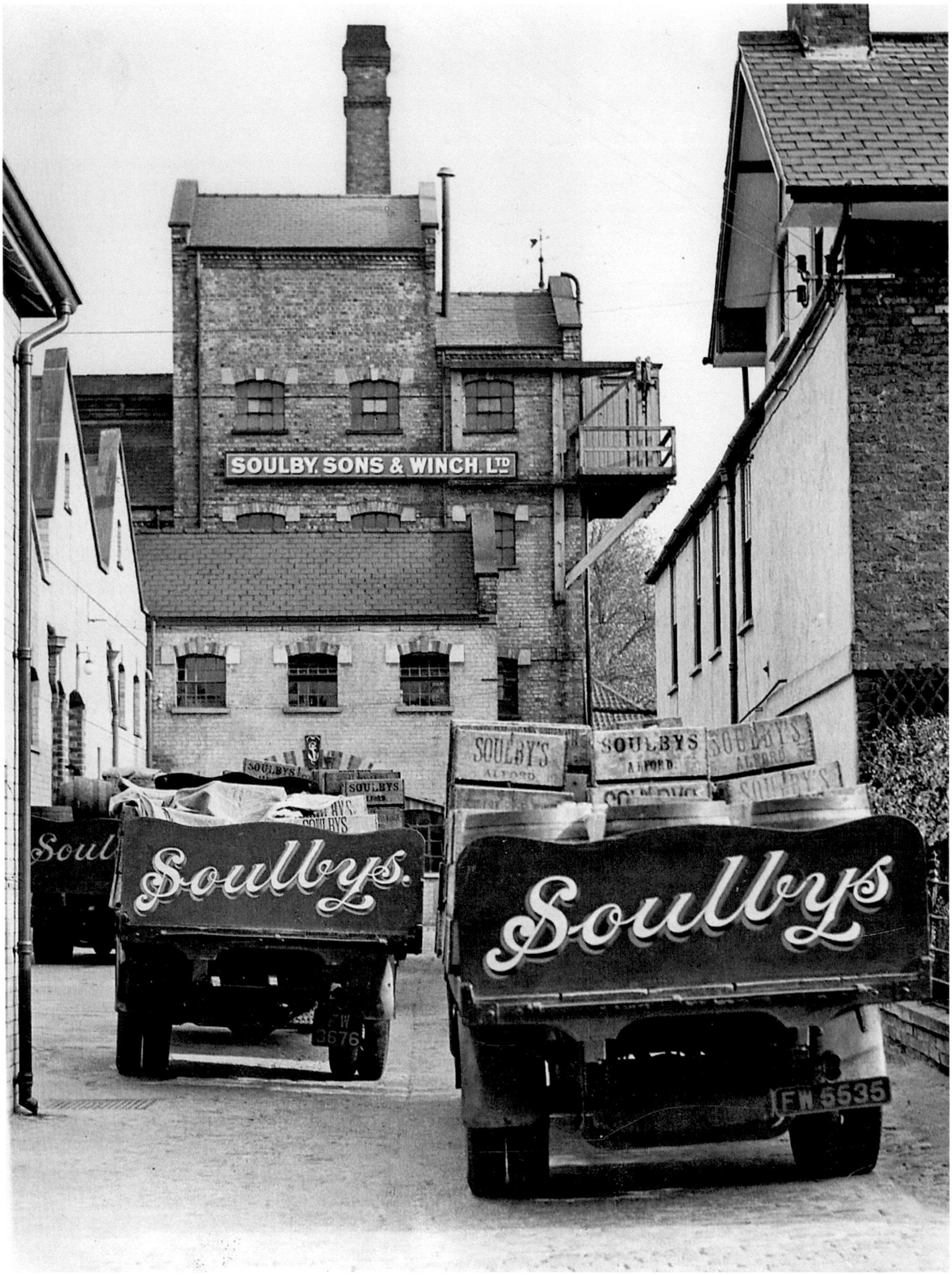

A charming view looking into the brewery from West Street in the late 1940s, showing most of Soulby's fleet of Bedford drays. The keystone with its anchor logo above the middle vehicle survives in the town's museum. (AC collection)

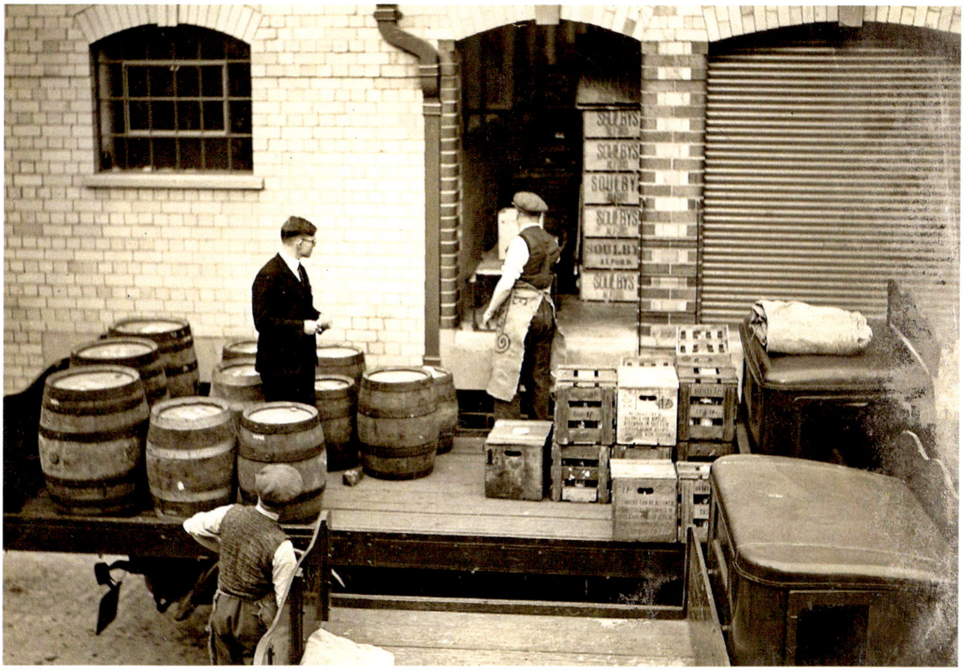

Checking a load before it leaves the brewery around 1950. It's believed that the man to the left is a young Ken Ketteringham, who spent most of his working life in brewing, moving to Mowbray's at Grantham when Green's bought Soulby's. (AC collection)

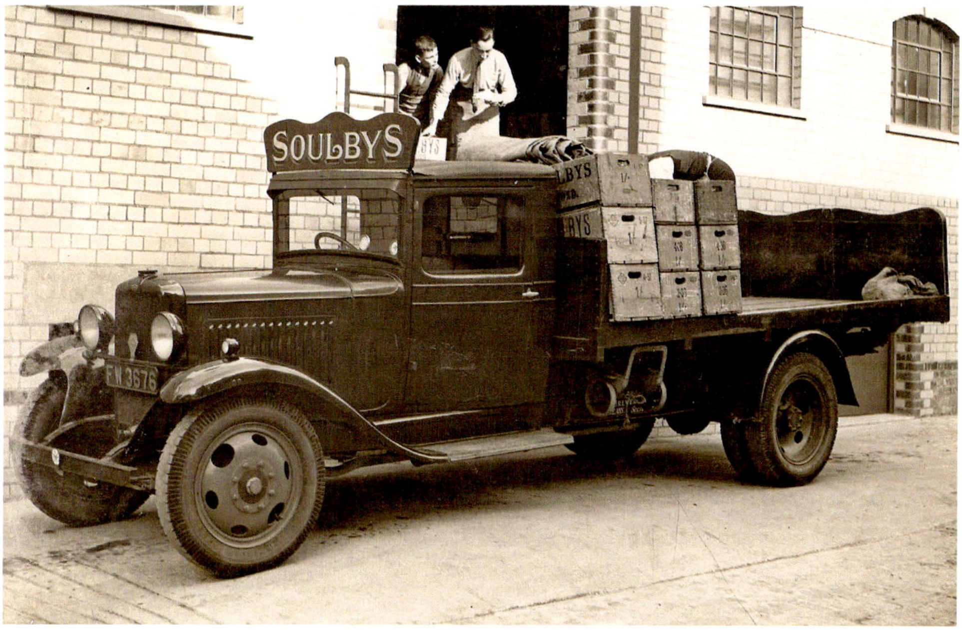

More deliveries being prepared at the brewery with a Bedford lorry dating from around 1932 ready for the road. The name of Cyril Jay, company secretary, is just visible painted on the chassis underneath the barrels. (AC collection)

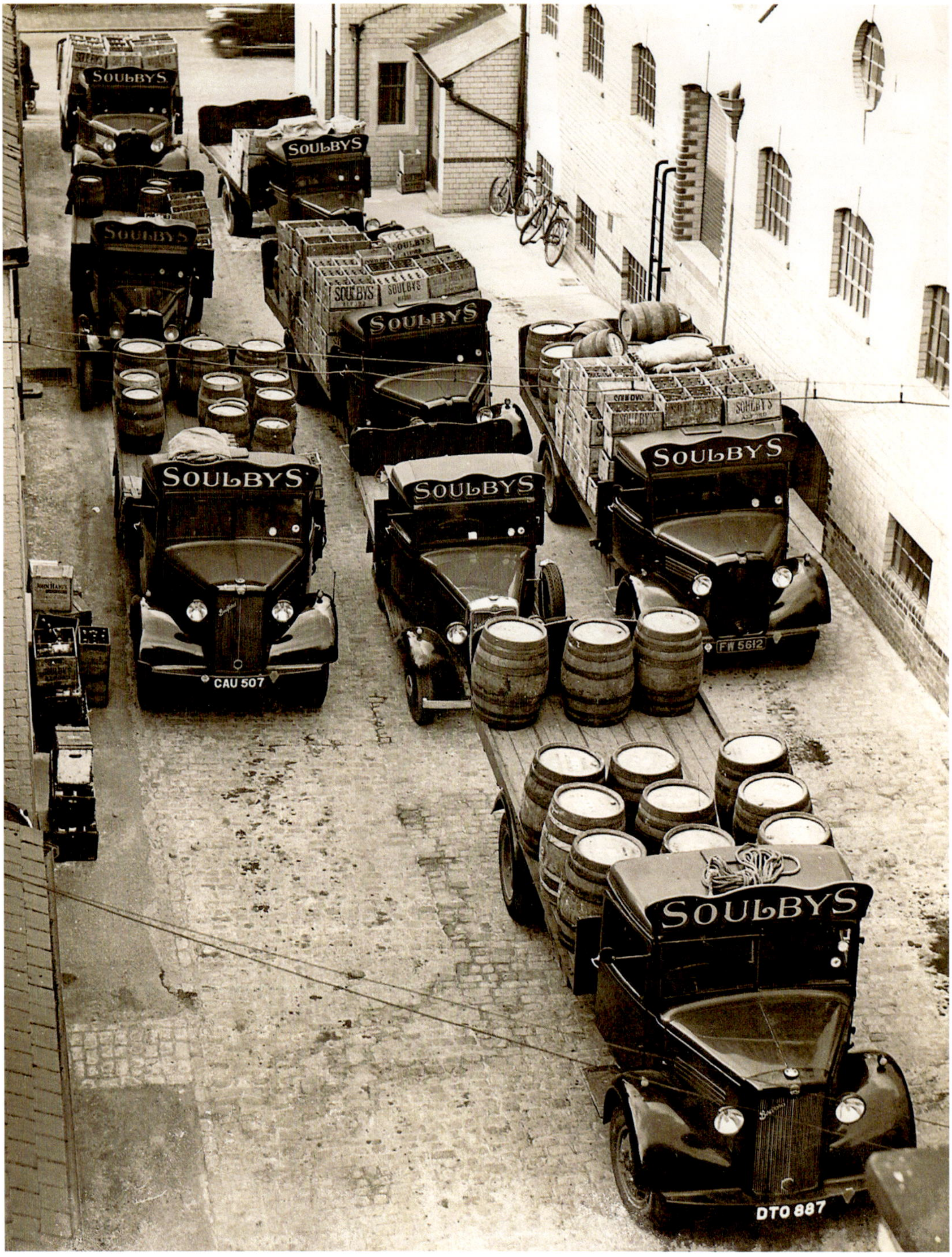

The fleet's in: eight drays, probably all Bedfords, seen from the brewery tower around 1945. Most are stacked with barrels and capped bottles, albeit somewhat precariously in some cases; let's hope an emergency stop wasn't needed. At least two of the lorries have Nottinghamshire registrations, indicating a life before Soulby's. (AC collection)

sold. The deal raised many eyebrows locally: Green's were a wholly unknown entity, based over 100 miles to the south, but they had embarked on a very rapid expansion programme, growing quickly to 600 houses, and were keen to add more. In February 1952 they bought Mowbray's of Grantham, as explained earlier. Whilst Mowbray's brewery was even older, and occupied a more restricted site, its location close to the A1 and a main line railway meant it was much better placed to serve a potentially wide selection of pubs. Rumours circulated almost immediately that Alford would be closed, which quickly proved correct, although the maltings remained in production.

The closure of the Anchor Brewery in September 1952 was a huge blow to its sixty-two employees and their families. For some of the administrative and supervisory staff, including managing director Colonel Cyril Jay and company secretary Ken Ketteringham, transfers were offered to Grantham, 50 miles away, but many were not.

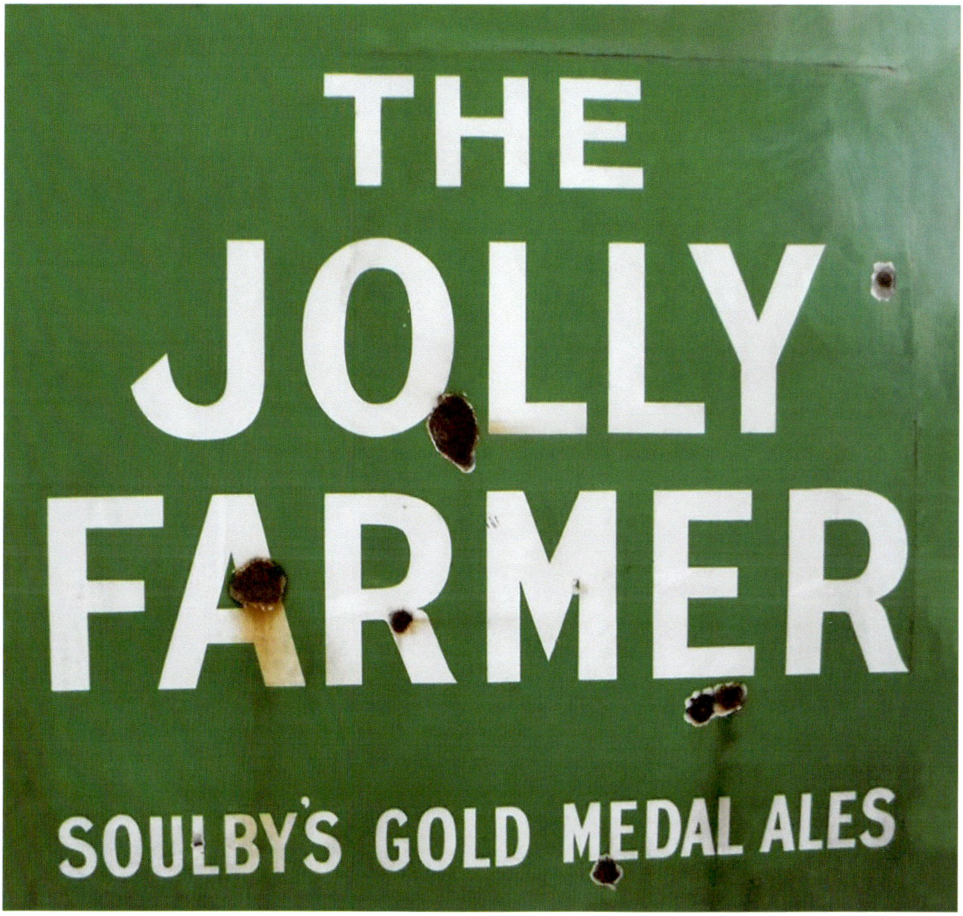

Soulby's often used green enamel signs for their pubs and little other signage. The 'Jolly Farmer' was just one example, a roadside beerhouse at Haltoft End near Boston, one of many small pubs on the A52 towards Skegness. It closed in the 1960s, but the sign has survived. Another example can be seen just inside the doorway of the Eagle in Boston. (AC)

Better news came when a Leicester hosiery firm announced it had bought the brewery premises for conversion into a nylon clothing factory. The brewing plant was stripped out, to be replaced with clothing machinery, although progress was slow, to the extent that the old brewery could be used as temporary housing for east coast flood victims in February 1953. Grand plans to employ over 300 workers at the factory were never realised, although some former Soulby staff were taken on.

Green's came to regret buying Soulby's. Running country pubs with declining custom was a drain on every brewery, and Green's had started closing the most hopeless cases, but they had added to their problems by creating a very long supply chain into rural east and south Lincolnshire. In 1956, around sixty former Soulby pubs were sold, half – the most northerly ones – going to Hewitt's of Grimsby and the remainder to Batemans of Wainfleet. All of the remaining pubs were sold off in 1957 and 1958, most going to Steward and Patteson of Norwich. Hewitt's also bought the Alford maltings, which finally closed in 1967.

Several of the old Soulby pubs remain open today, mostly in the hands of Batemans, and some 'ghost sign' reminders on the walls of former pubs can still be found, getting more faded with every year. The Anchor Brewery and its maltings are now sadly long gone, a victim of increasing dereliction and damage after the premises ceased to be used as a clothing factory in 2001. Planning permission exists to build housing on the site; originally, the maltings were to be retained, but their condition was such that demolition was preferred.

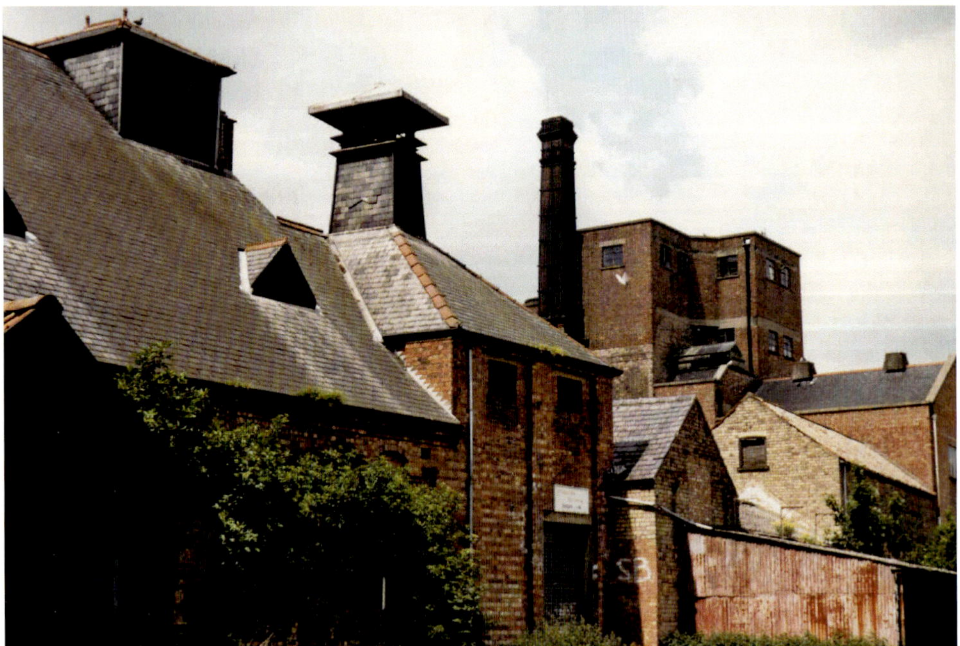

Soulby's brewery complex stood for several years after it was finally abandoned. On the left is the elegant maltings and to the right is the brewery, showing the inelegant brewery tower which was rebuilt after fire damage in 1945. (Brewery History Society)

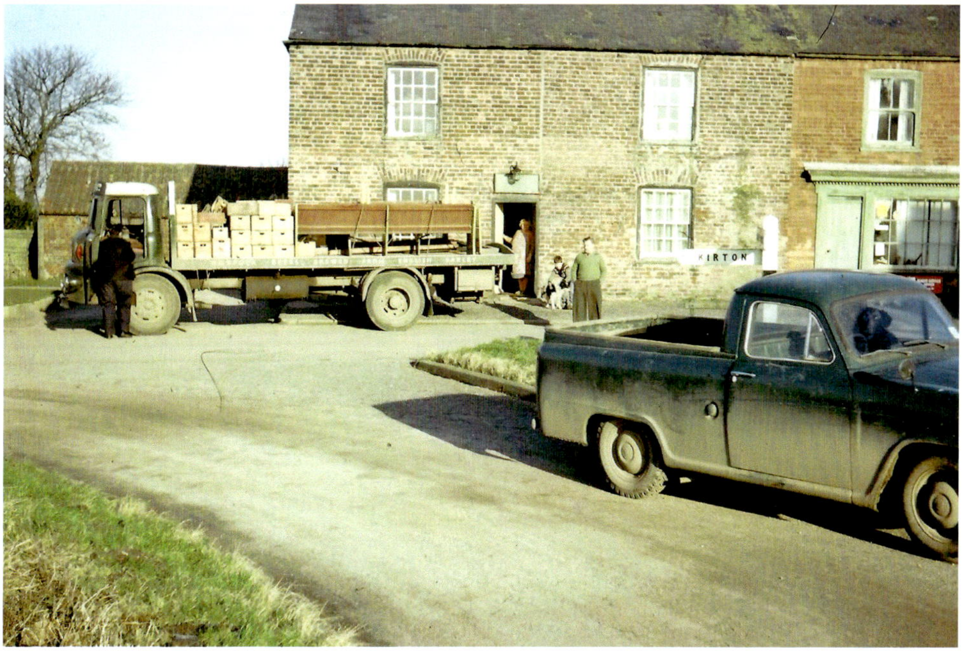

As social patterns changed in the 1950s and 1960s, scenes like this happened more often. Mrs Violet Willerton, the landlady of the Boat and Gun at Kirton Skeldyke, a mile from the Wash shoreline, looks on from her doorway with her grandson as Steward & Patteson's lorry collects stock and furniture on a sunny January day in 1967. The pub closed on New Year's Day 1966. Owned by Soulby's for many years, it was bought by S&P in 1957. (Paul Holmes)

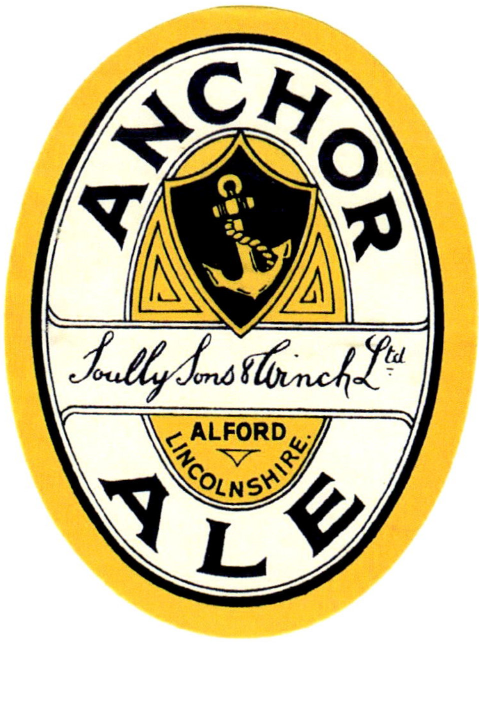

A beer label for Soulby's Anchor Ale, complete with nautical logo. The maritime theme reflects Alford's closeness to the North Sea, 7 miles from the town. (AC)

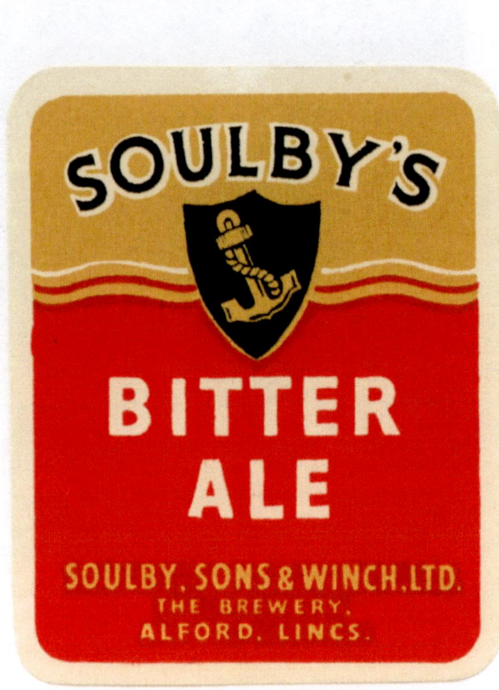

Soulby's Bitter Ale bottle label. This clean style, still with the anchor logo, was the brewery's final design, used across their beer range. (AC)

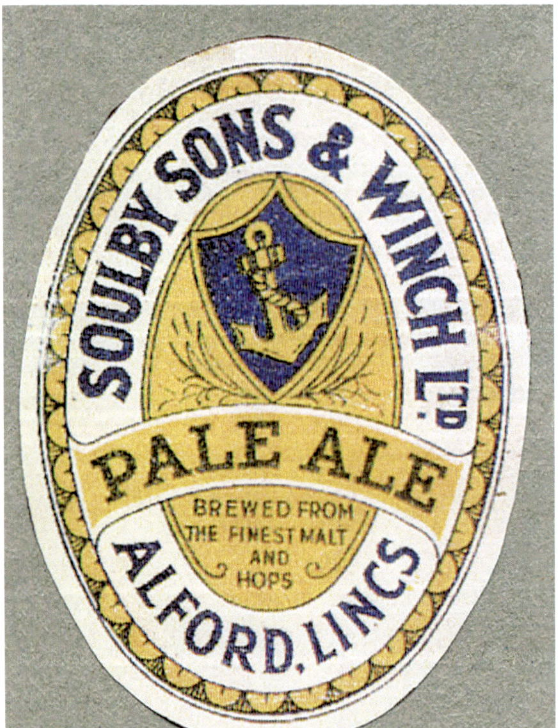

Soulby's Pale Ale label. The assurance that the beer was 'brewed from the finest malt and hops' was a typical quality assertion of the early twentieth century. (AC)

7

8 Sail of Heckington

As a respite from breweries of the past, let's switch to the present, and take a look at a couple of the county's successful modern breweries. First up is 8 Sail, a distinctive microbrewery trading from a heritage site in the village of Heckington, just to the east of Sleaford. Founded back in 2010, it's made a name for itself by combining a strong sense of community service with an appreciation of traditional brewing, but with a clear focus on the future.

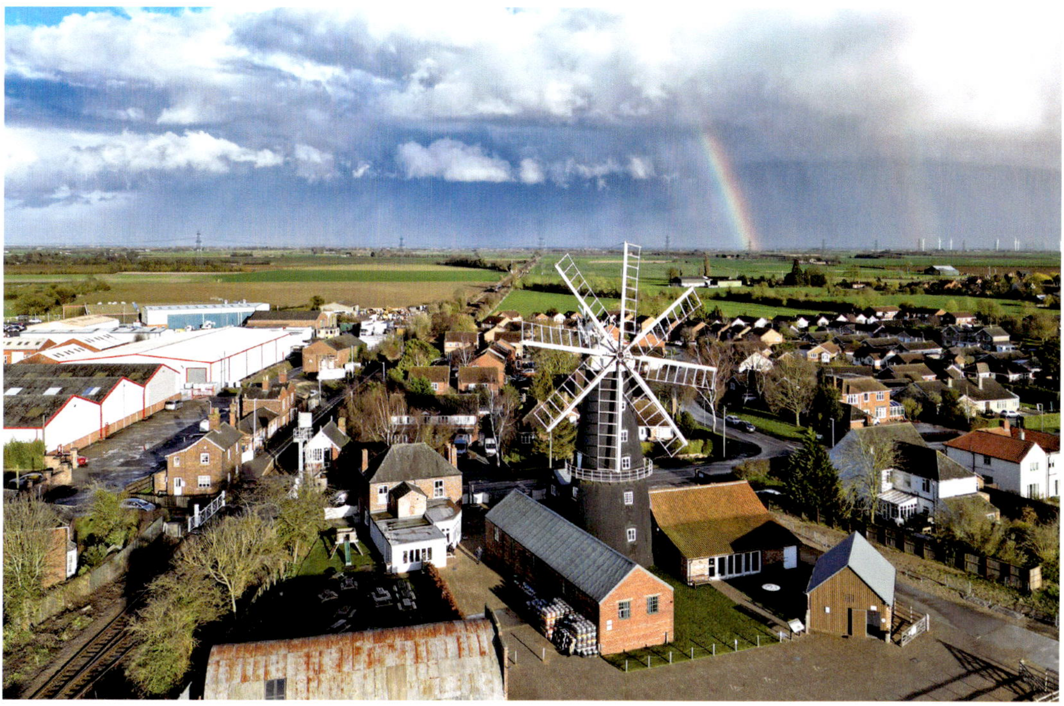

Heckington's eight-sail windmill towers over the brewery, which is the building immediately to its left in this captivating aerial image, captured on a day of dramatic weather in March 2023. (Rob Higgins)

Increasingly disenchanted with his career as a primary school head teacher, 8 Sail founder Tony Pygott had begun to consider his options – or, as he puts it, thirty years of teaching drove him to drink. He'd always been interested in home brewing and brewery history, and despite the well-meaning warnings of doom ('don't do it!') from local brewers, he made his decision. He didn't jump in totally blind, though: he'd undertaken a couple of brewing courses and sensibly retained a part-time teaching role, just in case.

Tony's search for a possible brewery site led him to some initially promising locations, but they all had their drawbacks. So, when Martyn Pocklington, a hockey team-mate, suggested creating a brewery out of an old sack store on his family's land, Tony needed some convincing. The single-storey building was derelict, with no doors, windows, drainage or services of any kind, a hole in the roof and a large number of resident pigeons. However, the location wasn't bad at all: Heckington is a large village, close to main roads and towering over the site was a very distinctive eight-sailed windmill which would lend the new brewery its name. Almost 200 years old, the Grade II structure was built in 1830 with five sails, with an upgrade to eight sixty years later – in fact, as Tony points out, it's the only working eight-sailed mill in Western Europe. Once again, a decision was made, and 8 Sail was born.

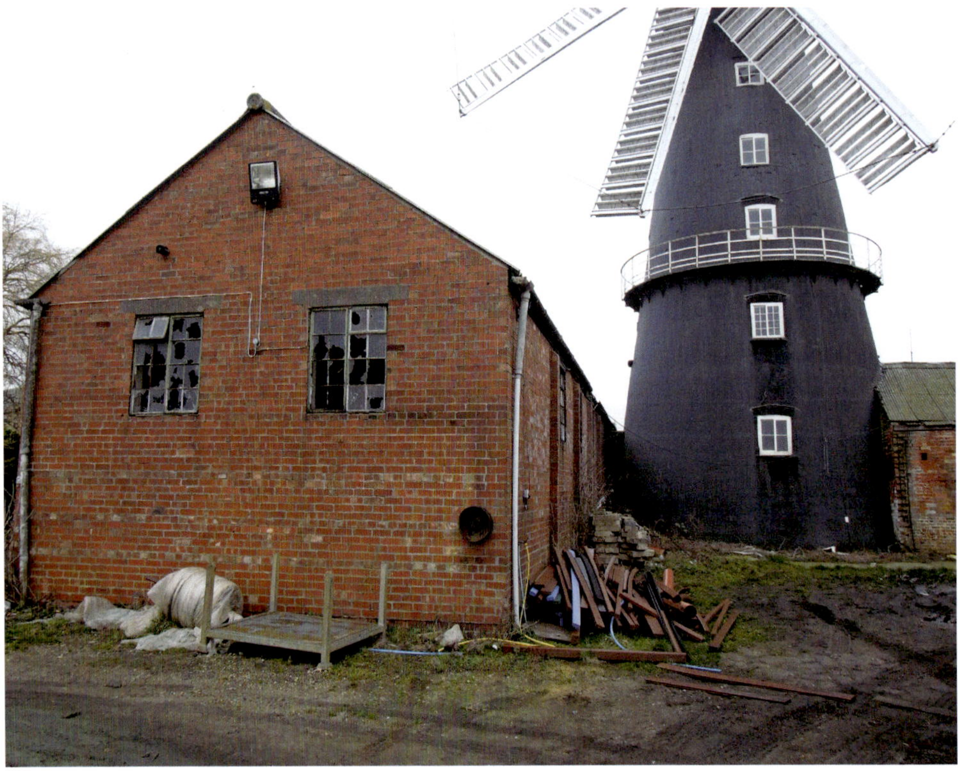

This is how the brewery building looked in 2009, with much work needed inside and out. The distinctive presence of the eight-sailed mill awaits developments with interest. (Tony Pygott)

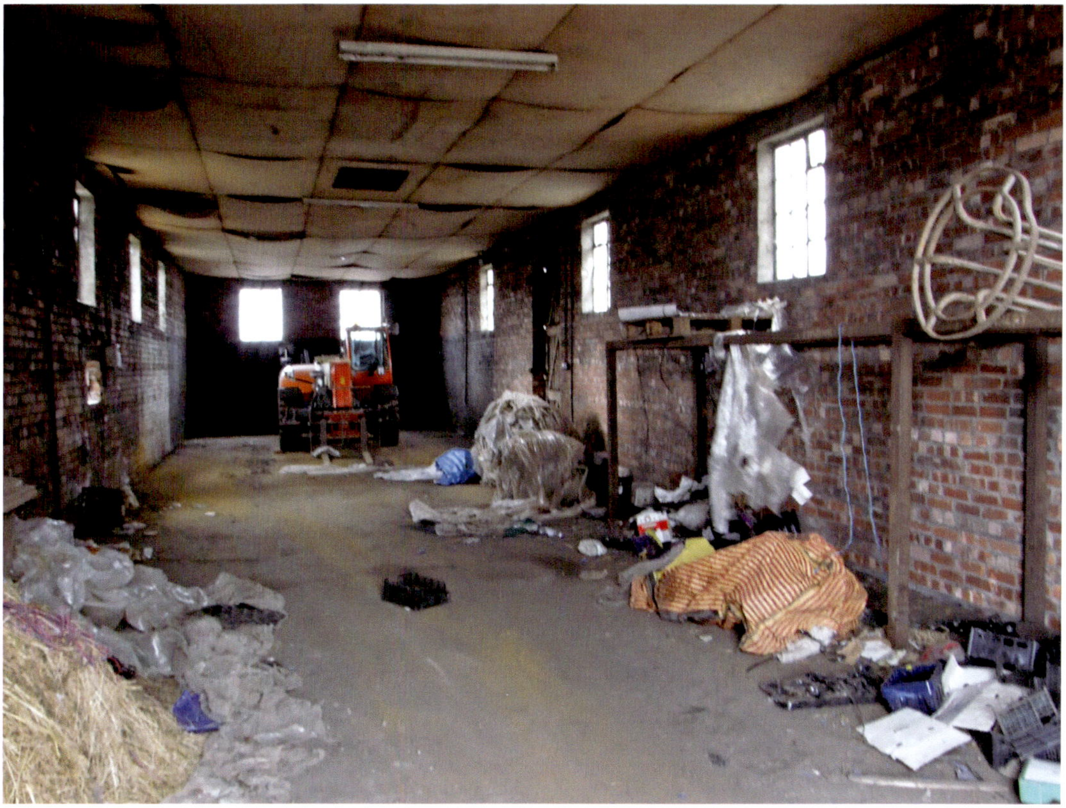

Inside, work had started to clear the building, with the suspended ceiling about to be removed. This view today is unrecognisable; the telehandler is standing in the area now taken up with the brewery tap. (Tony Pygott)

It took two years to get the building into shape, with the help of the Pocklington family and a supportive local district council. A five-and-a-half-barrel brewing kit with two fermenting vessels was installed and brewing got underway in 2010. The first three beers to emerge were a pale ale, called 8 Sail Ale, a traditional bitter called Merry Miller and Victorian Porter, a stronger dark beer that continues to win a string of awards and remains Tony's favourite to this day. I ask Tony how he comes up with new ideas for beers. 'Once you start brewing the urge to try different beer styles is very compelling', he says. 'The challenge is to brew the best you can, so you read through old recipes, take advice from suppliers and do your market research. But in the early years we aimed to brew beers we enjoyed drinking ourselves. So, we'd try out a new beer to find if it was popular, and if it was, keep brewing it the same way.'

One of the biggest challenges was getting 8 Sail beers into pubs. Tony always had confidence in his own brews, but he agrees you have to earn a presence on pub bars. 8 Sail had no pubs, so he toured the local free houses, many of whom were interested in something new to offer their customers. 'That was one of the reasons we had a large range of beers' recalls Tony. 'We had a range of between ten and fifteen beers at the time – an advantage of being small is that you have that flexibility – and I think that variety helped us get the beer into pubs.' It took a while but eventually persistence and a belief in the products paid off and 8 Sail beers started to become more visible around the local drinking scene.

Two years later, Steve Doane became a partner in the business. He'd just retired from the RAF after thirty-one years, much of which had been spent flying as a radar operator in Nimrod early warning aircraft, so the move into brewing was every bit as much of a career change as Tony's had been. Steve brought a degree of military precision, and between them they improved consistency and gained a reputation for quality beers. It was Steve who introduced American-inspired beers like Old Colony – one of his ancestors, John Doane, had been one of the early settlers of New England. The year 2013 saw the first of many awards for cask and bottled beer, notably Old Colony, Millwright Mild and Victorian Porter – the last named went on to win a bronze medal from the Campaign for Real Ale in 2018. Many other cask and bottled beers have been recognised since, the most satisfying being the recent awards for 8 Sail's ventures into keg and canned beers.

Around this time the whole site was acquired and renovated by the Heckington Windmill Trust. Their miller, Jim Bailey, approached 8 Sail with the idea of milling grain for brewing. National concern Bairds Malt, despite being sceptical, got involved with the project, agreeing to supply malted barley to Jim, whose confidence was rewarded with brewable grain. The ensuing collaboration, Rolling Stone, was launched with the help of Britain's first Beer Sommelier, Nigel Sadler. The information went out worldwide through milling and brewing guilds. It was clear that nobody else had tried the same thing, so 8 Sail were totally unique in calling Rolling Stone 'stoneground beer'.

Steve spent seven years with the business, retiring in 2019, a few months before the Covid-19 pandemic hit. The ensuing lockdowns badly affected the hospitality trade and, as Tony points out, the breweries that tended to struggle were those tucked away on industrial estates with no brewery tap or clear supply line into shops. With an existing retail trade, 8 Sail were better placed to ride the storm; the flexibility of a smaller business meant they could divert more production into bottling. But, as Tony notes, they needed some help. 'We were massively indebted to our regulars and locals who not only bought bottles but were able to take draught beer home as well. It meant we were able to break even – just.' During those difficult years, Tony was assisted by Lucy Irons, who moved over to brewing from a role at Anglian Water (there is some read across!) and apprentice brewer Callum.

From the start there had been a bar on the site, a rudimentary one at first, but the concept developed into a Victorian-themed taproom. Wooden panels from an old London pub were rescued from Germany, to where they'd been briefly exported, and married to a bar top which had previously been part of the science lab at a Nuneaton school – an unlikely pairing, but it works. Redundant church pews and cast-iron tables complete the look. The bar remains very popular to this day. It's open seven days a week, unlike most local pubs, and on the rainy Thursday lunchtime of my visit, there was a steady stream of customers.

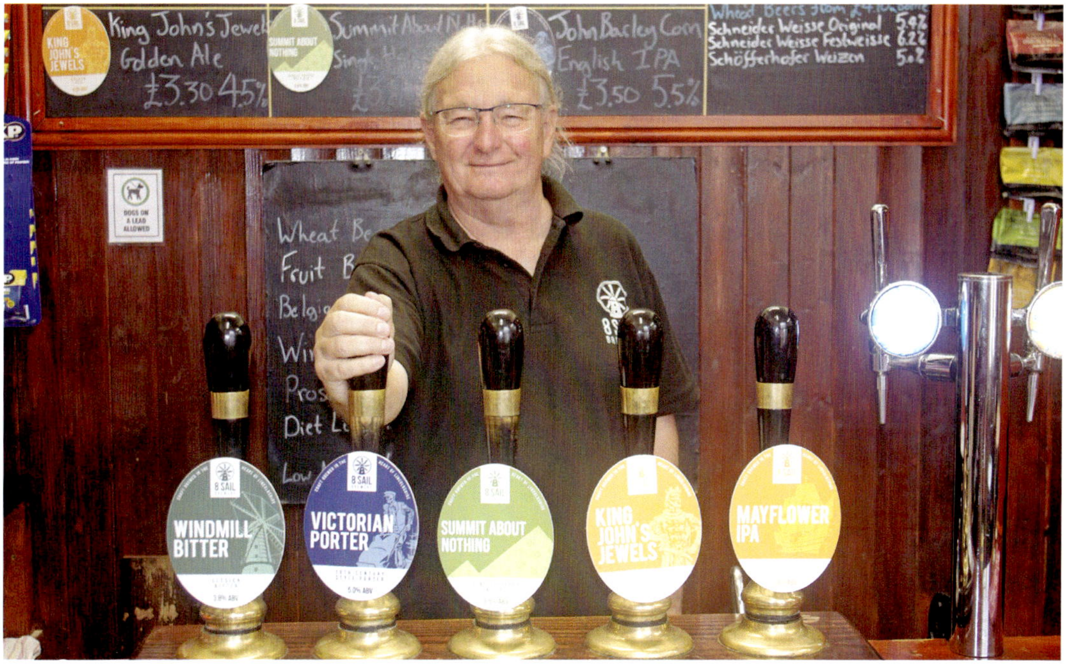

Founder Tony Pygott shows off the range of five hand-pumped beers available in the brewery tap on the day I visited in July 2024. (AC)

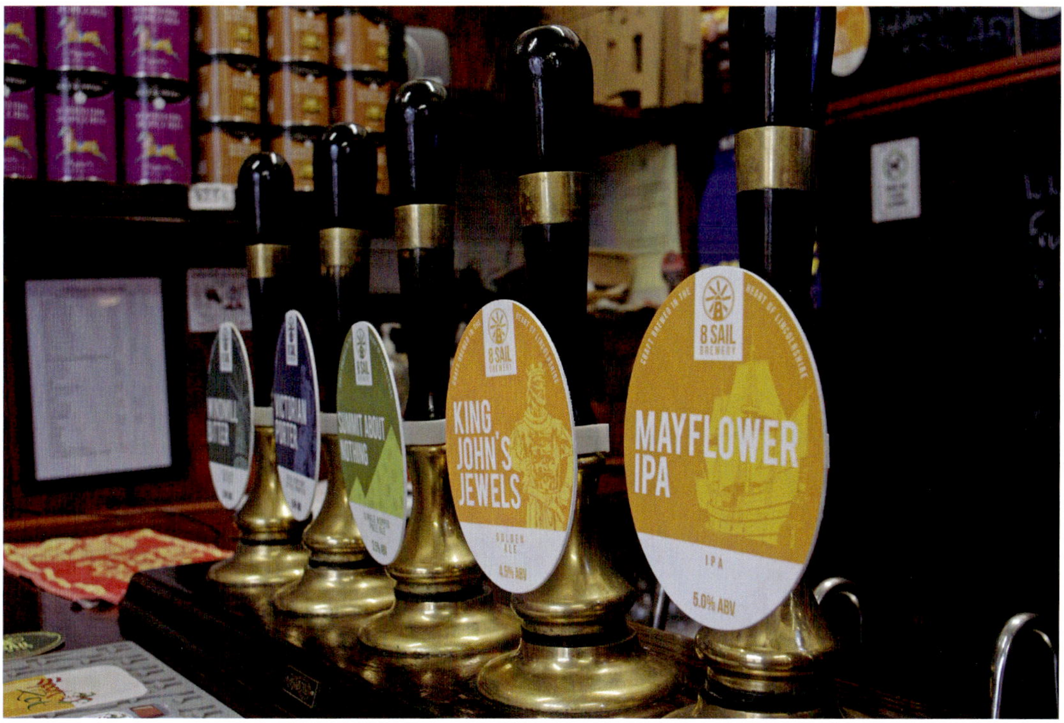

Five of 8 Sail's range of real ales await customers at the brewery tap. (AC)

When he started 8 Sail, Tony was keen to have a point of difference from the many other 'craft brewers' which were springing up in the late noughties, so as well as brewing cask ale from the outset, most of the range was also available in bottles, with a crucial difference: they were, and still are, what's called bottle conditioned. In other words, the beer undergoes a secondary fermentation in the bottle due to the presence of live yeast in the bottle which reacts with added sugars. This adds further character to the beer which remains 'live' and naturally carbonated, unlike the fizziness of conventional bottled and canned beers. It also tapped into the beer aficionado market quite effectively, many of whom fondly remembered bottle conditioned heritage beers of the past such as Worthington White Shield, once widely available but now defunct after many fitful years of occasional contract brewing. It's a process that some other brewers still follow, such as Thornbridge, whose nationally recognisable Jaipur IPA and more of its range can be had in bottle conditioned form.

The problem with bottle conditioning happens when the drinker removes the crown cap. You have to be careful when pouring to leave the bottom layer of yeast in the bottle, otherwise your glass will be full of cloudy beer. It's easily avoided with some judicious slanting of the specially shaped bottle, but it means that selling into the café and restaurant trade is a non-starter. You can't be constantly explaining to customers about special dispensing requirements, an issue which was brought into strong focus during the pandemic. So whilst 8 Sail won't abandon bottle conditioned, it's moving into other formats, starting with canned beer. A brace of new brews are being marketed under a new name, Hop Fen Brewing: Sky Lark is an American amber ale, whilst Windy Corner is a pale lager. Canning has some advantages over bottles: they are cheaper and lighter, but more importantly a can with an appealing label just looks better and more stylish than the traditional brown beer bottle. Kegging and carbonated bottled beers are other areas of interest, driven by the enthusiasm of new head brewer Paddy Bauwens, who brought valuable experience from breweries and hospitality in South Africa and the Netherlands. Many of his new recipes are superb, enthuses Tony.

There's been a design revamp too, with pump clips and bottle labels given an overhaul by marketing agency Harper Creative. The current brewing capacity is 2,000 barrels a year, and although 8 Sail doesn't brew quite that number, they do supply over 200 outlets mostly across Lincolnshire, although they also venture into Peterborough, Newark and Nottingham. They've also done their bit to support rural pubs, which were badly affected in the Covid lockdowns. The leases of two Fenland pubs, the Old Ship at Pointon and the Brass Windmill at Helpringham, were taken on to prevent their closure. 8 Sail managed to stabilise both businesses, passing on the leases to people who want to keep them going.

The most exciting development remains on the stocks. Tony would like to move most of the brewing operations to a new site a few miles away at Burton Pedwardine, although he'll still retain some capacity at Heckington, especially as the bar remains highly popular. It would streamline the business and allow storage at one site rather than spread across three rented locations, as well as provide space for future expansion if needed. The project has been paused after a disastrous partnering with a brewery installation company that almost crippled the business and forced them to instal a replacement second-hand brewing plant. But it's definitely something to watch for the future.

Reflecting on his fifteen years in brewing with 8 Sail, Tony enthuses about the creativity of the smaller brewers who have emerged in that period. Many have championed more niche brewing styles and flavours which have gained drinkers' interest and attention, and in turn given brewers more confidence to be even more experimental. Although it's led to some outlandish-tasting beers, it's placed the small guys ahead of the big corporates, who still control over 90 per cent of the drink trade. By nature more risk averse, they can flex their muscles by buying up small brewers who interest them, and offer cash incentives to the free trade to take their beers. Tony's hopeful for the future of smaller breweries, though. He reckons the way forward is to maintain strong community relationships with the pubs and farm shops 8 Sail supplies, making sure that they produce brews that customers enjoy and appreciate. He deeply regrets the decline of the traditional family brewer. 'So many have been taken over and well-loved beers have disappeared or become bland parodies of what they once were. All of them had tied estates which were their strength, though changing demographics have really impacted their viability.' He celebrates those that have survived.

He remains an optimist for the future. 'The last few years have cemented the place of the great British pub in people's minds. There was a real desire to get out and socialise, and enjoy the unique atmosphere of a pub. I think that's something that will remain for a long time yet.'

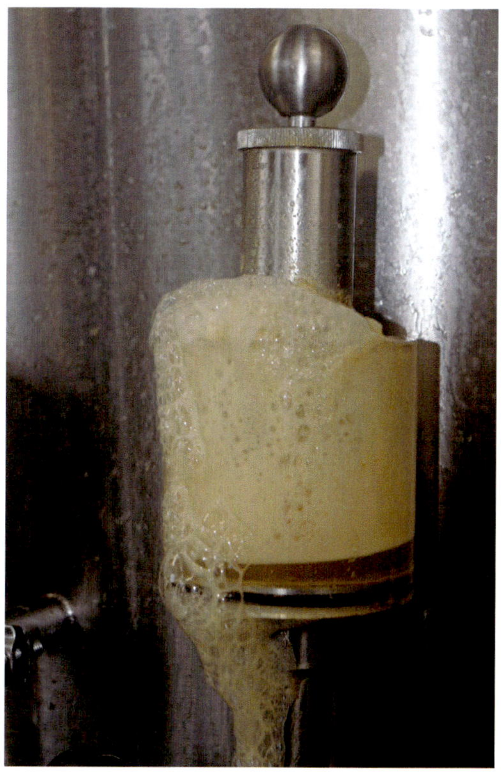

Overflow at the brewery. (AC)

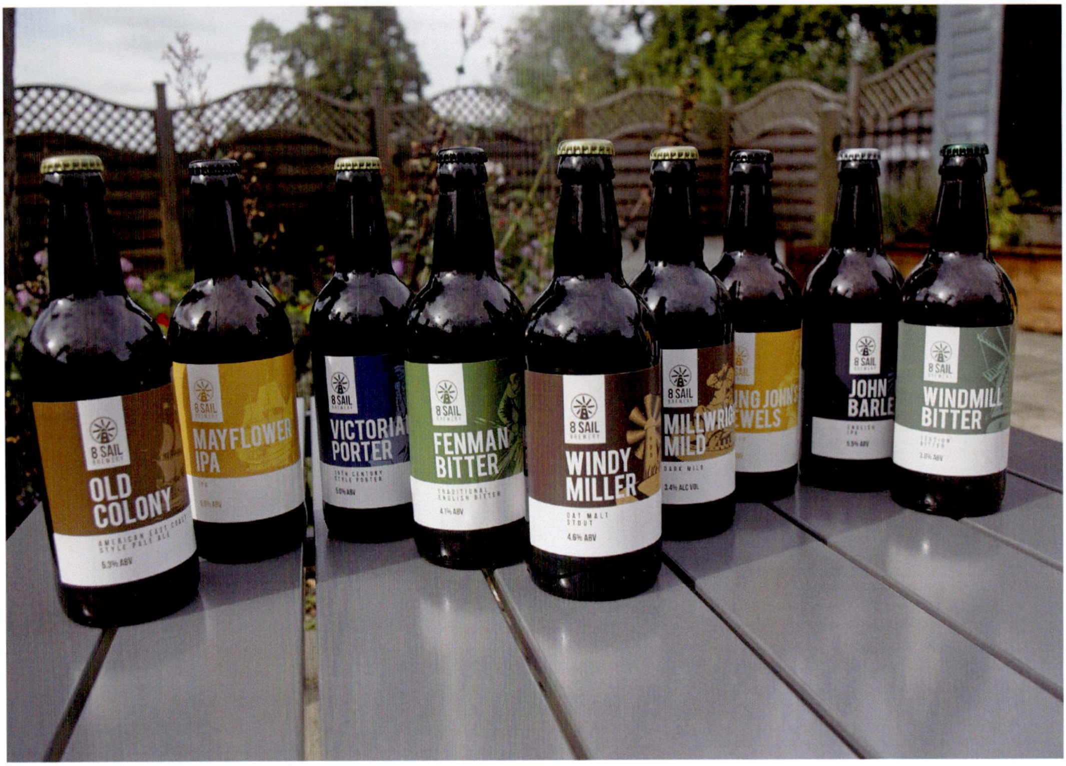

Most of 8 Sail's attractively branded bottled beers on display. (AC)

8

Ferry Ales of Fiskerton

For several weeks early in 2025 it seemed that one of Lincolnshire's best regarded microbreweries, Ferry Ales Brewery (or FAB), would be another casualty, buffeted by the harsh trading conditions across the hospitality sector and the hard physical work that brewing entails. Fortunately, a solution is at hand which will preserve their distinctive ales – a happy coincidence of right place, right time – but it was a close-run thing.

Situated at Ferry Hill Farm near Lincoln, Ferry may be unique: a brewery founded on a train journey. John Cussons and Mike Holman got talking on their commute into London and Mike casually mentioned he was getting fed up of the travelling and might start a microbrewery instead. John felt much the same way about the train. After a career in senior roles across professional service firms, working around the world, John and his wife had set down roots on an old smallholding close to the River Witham just east of Fiskerton village. He felt ready for something different, having dabbled in home brewing over the years, but had no thoughts of getting seriously into brewing until he started thinking about the old cow byre on his farm which just might be suitable. John and Mike began some serious research, including visits to Pheasantry Brewery near Retford, a microbrewery which had started a few years before in some converted farm buildings. Deciding to go ahead, they applied for a grant from LEADER, an EU rural development scheme, which provided £40,000 of the startup capital they needed. They chose an eye-catching name – Ferry Ales Brewery, which reflected the riverside location, but was easily shortened to FAB. It suggested great beer, but also for folk of a certain age harked back to a famous catchphrase from the space-age rescue missions of the TV series *Thunderbirds*.

John and Mike decided to go for the best-quality brewing kit they could afford, selecting ABUK of Lincoln as their equipment supplier. In October 2016 a five-barrel kit was installed in John's converted cowshed; they couldn't go any bigger due to the height of the roof. In those early days, cask beer was easily the best seller, but nonetheless the development grant was invested in a bottling machine. They weren't to know it at the time, but that was a decision which almost certainly stopped the business from going under just a few years later. Getting the new brewery's beer into

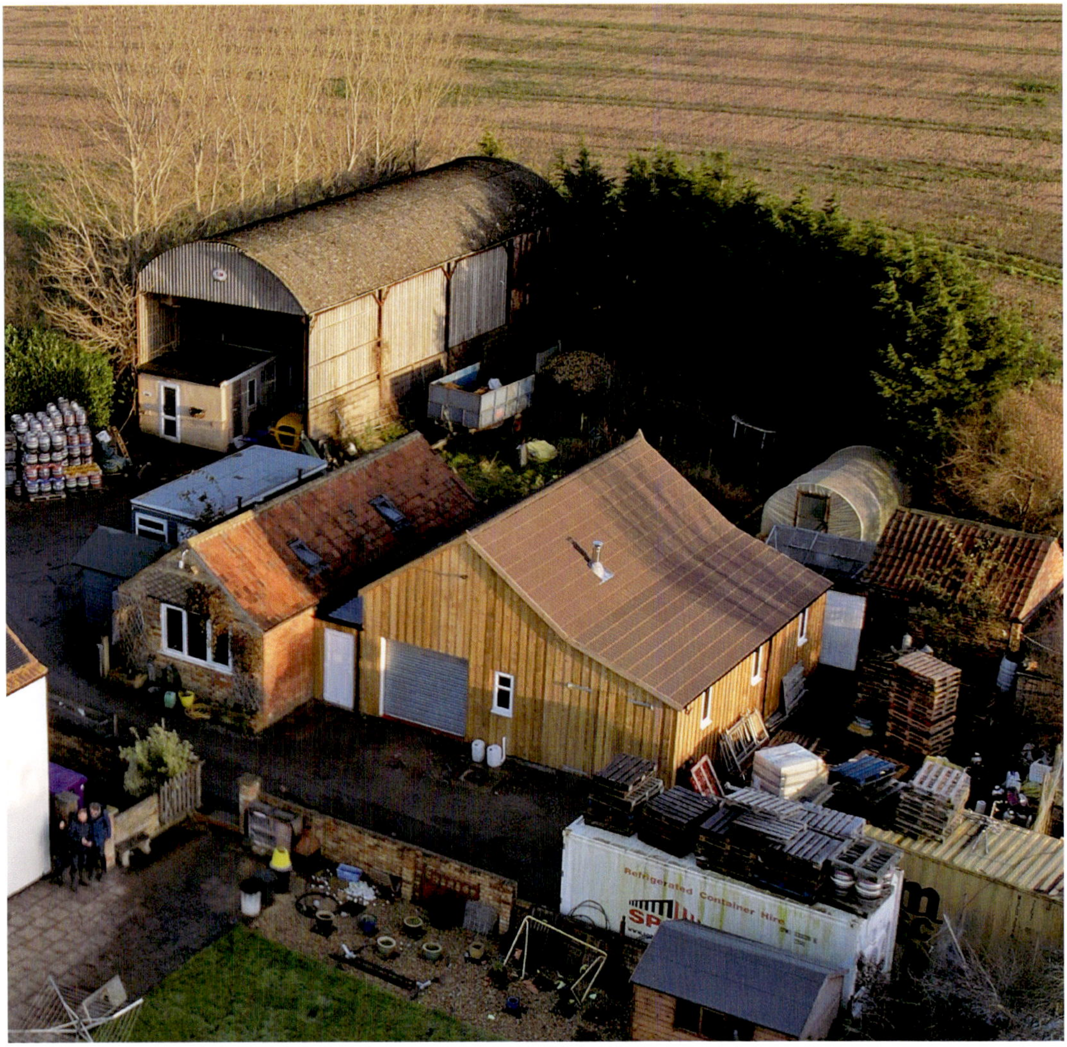

A drone shot of the brewery complex at Ferry Ales: the original cowshed is on the left, the wooden-clad extension to its right. (John Cussons)

pubs was hard going at first, echoing Tony Pygott's early experience at 8 Sail, but then they met James Cann, the operations director at an East Midlands pub company called the Ever So Sensible Group, whose suite of houses included four prominent outlets in and around Lincoln. That helped enormously, as did securing a contract to supply bottled beer to Lincolnshire Co-op's 100 food stores, raising FAB's profile across the county. Getting their beer to farm shops and independent retailers in Lincoln at an early stage was a further smart move. By December 2017, Ferry was doing well enough for John to leave his City job and devote his attentions full time to brewing.

Looking back now, John reckons he'd do things differently if he could have his time again, hindsight being what it is. 'I'd want to have my own outlet, whether a taproom or a small pub. And I'd find a venue in Lincoln and start there. Whilst we're

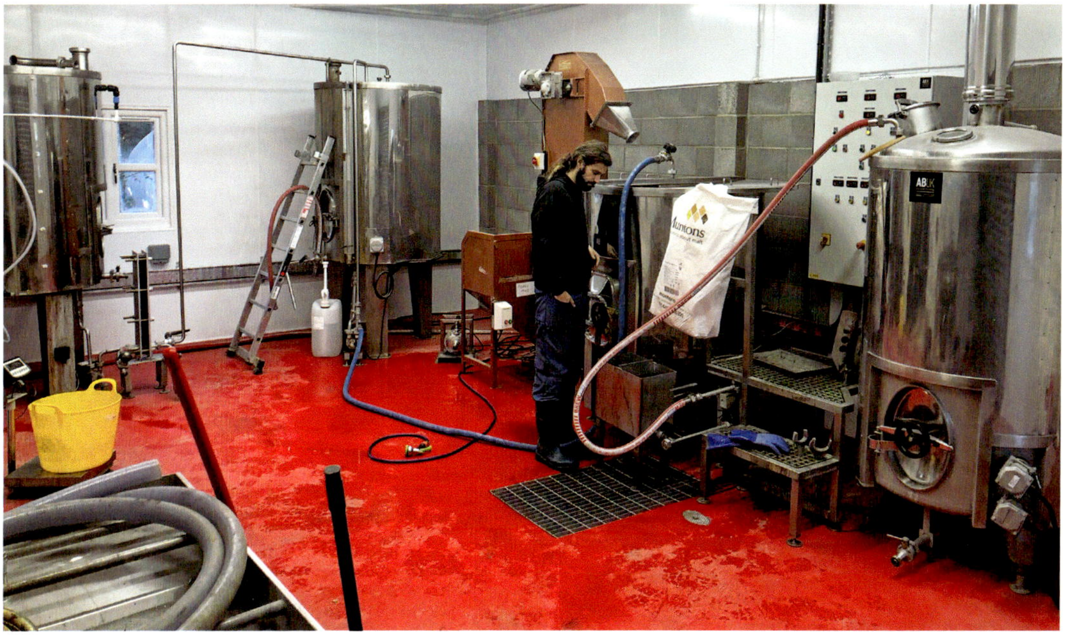

Brewing in progress in the new extension during 2022. The stainless-steel vessels, electronic controls and safety measures are a far cry from the Victorian-era breweries seen elsewhere in this book. (John Cussons)

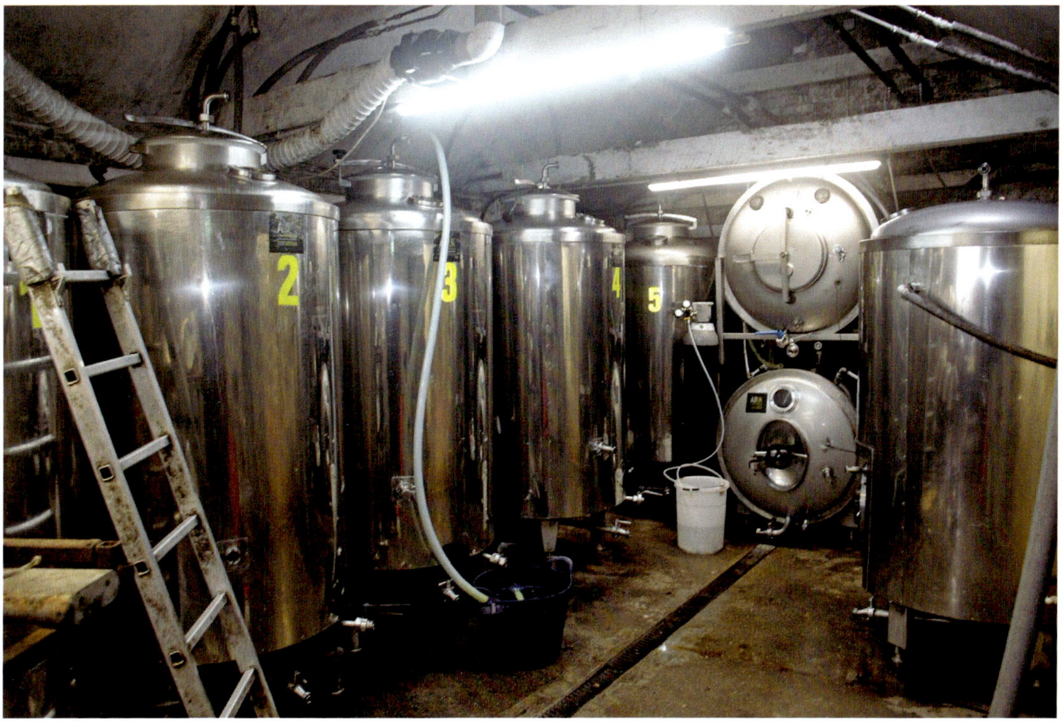

The original brewery building, given over to fermentation in 2025. (AC)

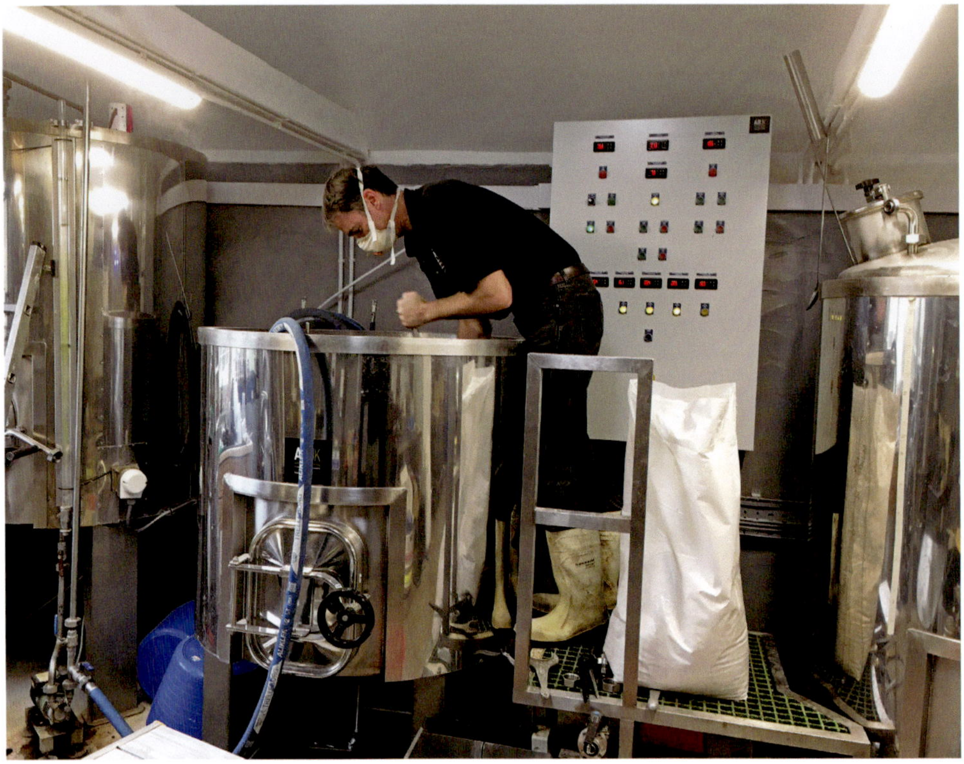

John Cussons mashing in at the start of the brewing process. (John Cussons)

fairly close to the city here, we're definitely not central. Your own place makes you self-sufficient.' He gives the example of Lincoln Green Brewing, who despite the name are based in Hucknall, over the Nottinghamshire border. They started off in a garage during 2012, four years before John and Mike, but now have eight pubs of their own and a brewery tap.

Although Mike decided to move on after three years, selling his shares to John and starting a pork scratchings business called Oink, the business continued to grow. But then Covid hit in March 2020. It was of course a terrible time: like every other brewer, they were very badly affected, with the pub trade disappearing overnight, and staff had to be furloughed. But the bottling plant came into its own as home sales soared. And as people built bars in their garden during lockdown summers, they bought 'beer in a box' – polypins of 5, 10 or 20 litres, for which FAB were, and still are, one of the top Google hits in the UK. Soon they were shipping as far away as Cornwall and Scotland. Production was down to John and his family; he reports that his children and mother-in-law were a huge help with bottling. John maintains what really saved the brewery, though, was local people. 'I put out a Facebook post asking if anyone wanted free delivery of our beers' he recalls. 'The response was overwhelming, and by the middle of the first lockdown we were sending three vans a week out of here.' They sold 100,000 bottled beers between March 2020 and March 2021, and John still offers free delivery locally within LN1 to LN10 postcodes as a continued thank you for that support.

As lockdowns started to ease, thankfully staff could return, and in August 2020 an assistant brewer, Jamie Morrison, was taken on. Soon the brewery was at capacity, and thoughts turned to expansion. John thought about moving production off site – being next to his home has its drawbacks – but he decided to go ahead with an extension behind the old cowshed, with a 20-foot container filling in whilst the additional space was constructed. The extension increased floor spare by 150 per cent, hosting the mash tun, kettle and liquor tanks, with the original building devoted to fermenting, conditioning and cask racking. The extension has also given the brewery a dedicated bottling space and the ability to brew up to four times a week. Production has increased to 6,000 pints a week and cask is back as a major seller, especially since it was realised that those polypins had great potential for pubs. They can be linked up to a handpump, thereby largely solving problems with beer going off before a standard barrel can be finished; it's helped several lower-volume pubs to bring back cask ale.

FAB continued to build up their events trade under the name FAB Bars. Starting in 2018, they'd got a toehold into providing bars at indoor and outdoor events, and since the end of Covid this side of the business has snowballed. Many big events want to deal with single suppliers for concessions, and FAB has some great relationships with the major local shows and Market Rasen racecourse, as well as serving smaller private events such as weddings and birthdays. They can offer a full range of drinks as well as FAB beers and are equipped with a cold room for proper cellar conditions, all under a marquee or tipi cover. The longest bar they've run so far was 10 metres long with twenty real ales on handpumps.

The range of beer has expanded well beyond the initial six, three chosen by Mike and three by John, based on the beer types that each of them liked, which they then tested on family, friends and pubgoers. The beers nearly all have Lincolnshire connections in their names, but John says the ones that resonate most with the public are those with an aviation theme. It started with 49 Squadron, one of those first six beers, an amber bitter which was named after those who flew from the former RAF Fiskerton. Stirring names such as Midnight Phantom and Mosquito (a plane flown by John's great-uncle) helped to boost sales, as did Just Jane, a brown session bitter named after the Lincolnshire Aviation Centre's Lancaster bomber: a portion of the sales support the veteran plane's bid to fly again. 'There's a strong interest in World War Two so these beers sell just as well outside the county – we've even had enquiries from overseas' says John.

Another beer which started out as a tie-up has gone on to be one of Ferry's biggest sellers. Lincoln Red was first brewed in 2021 in collaboration with the Lincoln Red Cattle Society as a celebration of the county's native beef cattle, which have a very distinctive deep cherry red coat. Based on an Irish red ale, its slightly sweeter taste made it an ideal beer to partner with roast beef, but its popularity quickly grew beyond farming circles. Ferry aim to use locally sourced ingredients as far as possible: Lincolnshire grows about a tenth of the UK's barley crop, a key beer ingredient in malted form, whilst some of their hops originally came from David Wykes at nearby Dunholme, who was the UK's most northerly commercial hop grower until Covid came along. David's hops featured in Harvest Gold, an award-winning citrussy golden ale

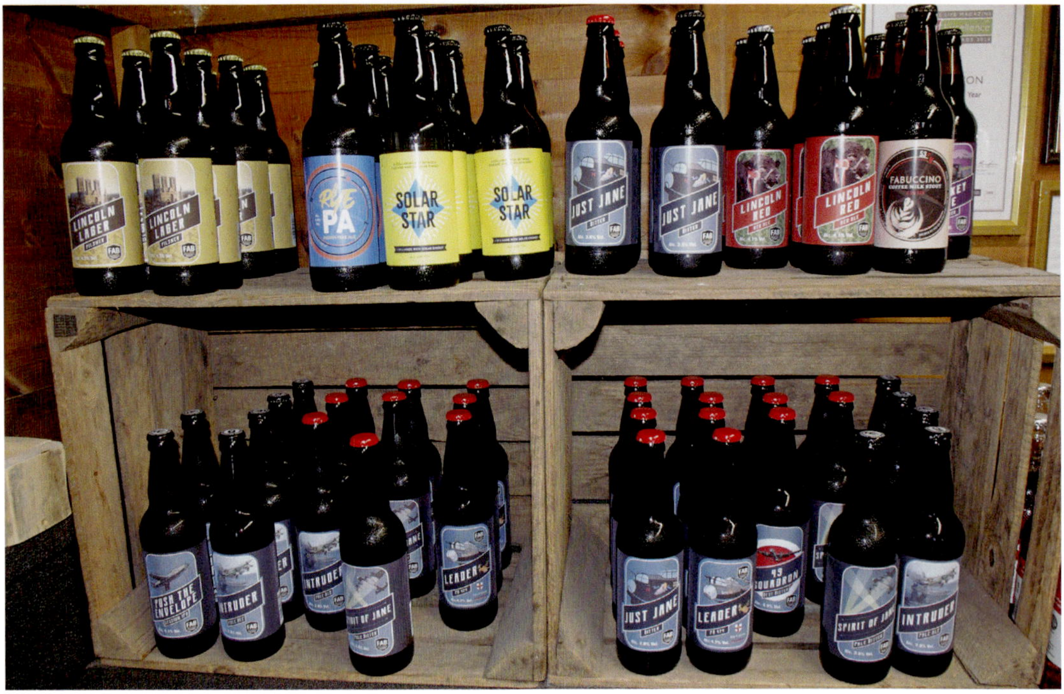

Most of Ferry's 2025 bottled beer range on display in the brewery shop. (AC)

first brewed in 2019 to celebrate the 150th anniversary of the Lincolnshire Agricultural Society. Ferry isn't afraid to try the more unusual, either, such as Fabuccino, a milk stout brewed with the addition of coffee supplied by long-established coffee house Stokes of Lincoln. And if you like, you can order a bottle with your name on it!

Brewing is hard work though, as many of the brewers we've looked at have found over the years. After nine years and something like 1 million pints brewed, John decided it was time to step back, and in January 2025 he made the difficult decision to close the brewery, which he scheduled for that March. However, a happy solution is at hand. Over at the Coach and Horses pub in nearby Billinghay, the beer range of publicans Luke and Catherine Mitchell had attracted much media attention. Osama Bin Lager, Kim Jong Ale and Orange Mussolini were just some of the names they'd come up with, the latter accompanied by an unflattering image of the US president. This tongue-in-cheek 'dictator' range, which supports various charities, led to a boom in sales as a result of some excitable headlines. Luke had already started to build a state-of-the-art brewery at Billinghay with the intention of moving production from his existing brewery in the West Midlands; however, he'd found that breaking into the Lincolnshire beer market was quite tough-going. Through a joint connection at Lincoln City Football club John and Luke realised there was an obvious solution: Mitchell Brewing would acquire the Ferry Ales business, thereby giving him a local brand and a local team. The deal was announced on the businesses' respective social media accounts in February 2025. For the time being, brewing will continue at Fiskerton, with production eventually moving to Billinghay. It's intended that FAB will

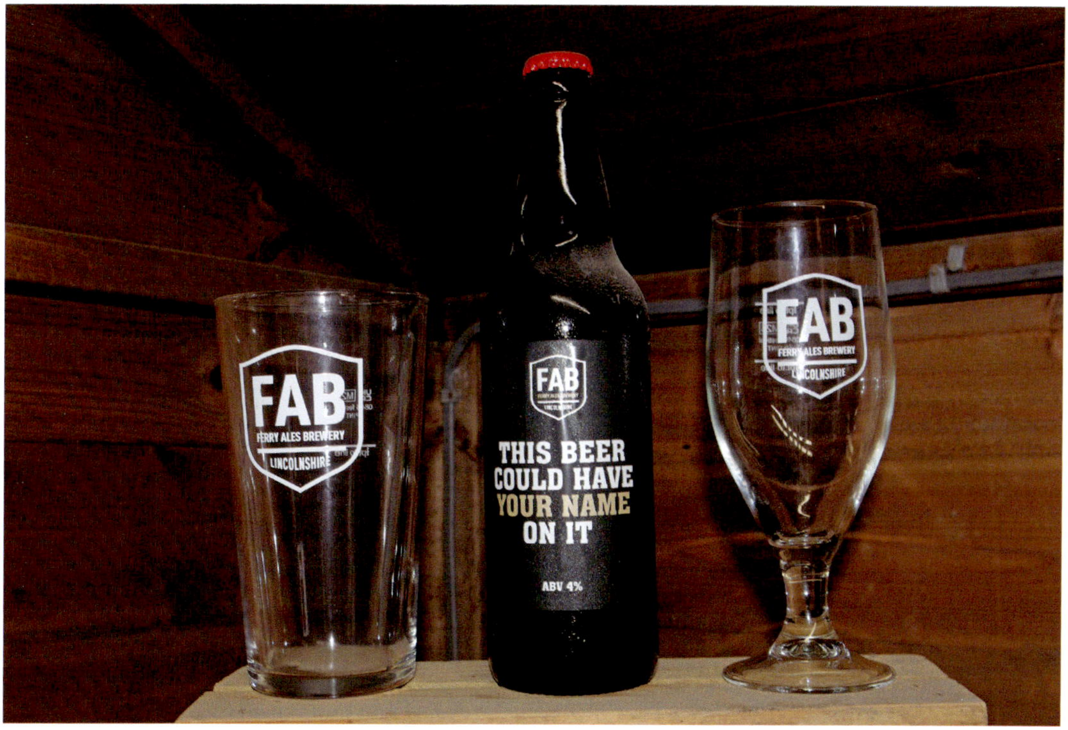

Some clever marketing by Ferry Ales. (AC)

remain as a separate brand to the Mitchell Brewing range, with FAB Bars remaining under John's control. But the most important thing is that this distinctive business and its employees survive to brew another day.

Further Reading

Aldabella, Pat and Barnard, Robert, 'Grimsby Breweries', *Brewery History*, No. 85 (Brewery History Society: 1996)

Aldabella, Pat and Barnard, Robert, 'North Lincolnshire breweries', *Brewery History*, No. 86 (Brewery History Society, 1996)

Birch, Neville, *Stamford's Industrial Past: An Untold Story* (Society for Lincolnshire History & Archaeology: 2021)

Brown, Mike, Farleigh, Ray and Smith, Ken (eds), *Brewers of the British Isles 1890 to 2021* (Brewery History Society: 2022)

Cartwright, Adam, *Good Honest Tales: 150 Years of Batemans Brewery* (Amberley Publishing: 2024)

Cartwright, Adam, 'Mowbray & Co. Ltd, Brewers of Grantham', *Lincolnshire History & Archaeology*, Vol. 49 (Society for Lincolnshire History & Archaeology: 2017)

Cartwright, Adam, 'Soames of Spalding', *Brewery History*, No. 157 (Brewery History Society: 2014)

Cartwright, Adam, 'Soulby, Sons & Winch Ltd', *Lincolnshire History & Archaeology*, Vol. 47 (Society for Lincolnshire History & Archaeology: 2014)

Dye, Geoff, '150 Years of the Melbourn Brothers' All Saints Brewery', www.breweryhistory.com

Gourvish, Terry and Wilson, Richard, *The British Brewing Industry 1830–1980* (Cambridge University Press: 1994)

Larn, Graham *Beer, Hope and Charity: Hewitt's Grimsby Brewery Remembered* (Century Zero Four: 2008)

Smith, Martin, *Stamford Pubs & Breweries* (Spiegl Press: 2006)

Walker, Andrew (ed.), *Pubs in Lincoln: A History* (The Survey of Lincoln: 2017)